Complete Biology for Cambridge IGCSE® Workbook

For the updated syllabus

Ron Pickering

Oxford excellence for Cambridge IGCSE®

Great Clarendon Street, Oxford, OX2 6DP, United Kingdom

Oxford University Press is a department of the University of Oxford. It furthers the University's objective of excellence in research, scholarship, and education by publishing worldwide. Oxford is a registered trade mark of Oxford University Press in the UK and in certain other countries

© Oxford University Press 2016

The moral rights of the authors have been asserted

First published in 2016

All rights reserved. No part of this publication may be reproduced, stored in a retrieval system, or transmitted, in any form or by any means, without the prior permission in writing of Oxford University Press, or as expressly permitted by law, by licence or under terms agreed with the appropriate reprographics rights organization. Enquiries concerning reproduction outside the scope of the above should be sent to the Rights Department, Oxford University Press, at the address above.

You must not circulate this work in any other form and you must impose this same condition on any acquirer

British Library Cataloguing in Publication Data
Data available

978-0-19-837464-0

9 10 8

Paper used in the production of this book is a natural, recyclable product made from wood grown in sustainable forests. The manufacturing process conforms to the environmental regulations of the country of origin.

Printed by CPI Group (UK) Ltd, Croydon CR0 4YY

Acknowledgements

®IGCSE is the registered trademark of Cambridge International Examinations.

The publishers would like to thank the following for permissions to use their photographs:

Cover image: Eduardo Rivero/Shutterstock; p12: Jiri Hera/Shutterstock; p107: Ron Pickering.

Artwork by Q2A Media Pvt. Ltd. and OUP.

Although we have made every effort to trace and contact all copyright holders before publication this has not been possible in all cases. If notified, the publisher will rectify any errors or omissions at the earliest opportunity.

Links to third party websites are provided by Oxford in good faith and for information only. Oxford disclaims any responsibility for the materials contained in any third party website referenced in this work.

Introduction

When using this workbook you will have the opportunity to develop the knowledge and skills that you need to do well in each of the papers in your IGCSE Biology examination.

The IGCSE syllabus explains that you will be tested in three different ways. These are called Assessment Objectives (AO for short). What these AOs mean to you in the examination is explained below:

Assessment Objective	What the syllabus calls these objectives	What this means in the examination
AO1	Knowledge with understanding	Questions which mainly test your recall (and understanding) of what you have learned. About 50% of the marks in the examination are for AO1.
AO2	Handling information and problem solving	Using what you have learned in unfamiliar situations. These questions often ask you to examine data in tables or graphs, or to carry out calculations. About 30% of the marks are for AO2.
AO3	Experimental skills and investigations	These are tested on the Practical Paper or the Alternative to Practical (20% of the total marks). However, the skills you develop in practising for these papers may well be valuable in handling questions on the theory papers.

Notice that the **recall** questions (AO1) only account for 50% of the marks – you need to show your skill in using these facts for the remaining 50% of the marks.

This workbook contains many exercises to help you to check your recall and to practise these skills. They will be similar to many of the questions you will actually see in your examination, so you will also be helped to develop the skill of working in an examination. In particular, you will find that many of the exercises cover factual material from different parts of the syllabus – exactly like the more difficult questions in the examination. Each worksheet contains an extension exercise to extend your learning beyond the syllabus and the examination. These include research and project tasks that will develop your scientific skills and understanding.

The answers to the questions are provided, so that you can assess your own performance. Be honest with yourself when checking the marks – you must not be more generous than an examiner would be! Your teacher will probably be able to help you to compare your performance with the expected standards. (Answers to extension questions are not provided.)

Practice may not make perfect, but it will certainly make better.

Good luck!
Ron Pickering

Contents

1a Characteristics and classification of living organisms

1.1	Life and living organisms	2
1.2	Using a key for garden birds	3
1.3	The life of plants	4
1.4	Invertebrates in woodland	5
1.5	Features of vertebrates	6

1b Cells and organisation

| 1.6 | Cells | 7 |
| 1.7 | The organisation of living organisms | 8 |

2a Diffusion and osmosis

| 2.1 | Movement in and out of cells: diffusion | 9 |
| 2.2 | Movement in and out of cells: osmosis | 10 |

2b Enzymes and biological molecules

2.3	Organic molecules	11
2.4	Testing for biochemicals	12
2.5	Enzymes	13
2.6	Enzyme experiments	14

2c Photosynthesis and plant nutrition

2.7	Photosynthesis	15
2.8	The rate of photosynthesis	16
2.9	Leaf structure and photosynthesis	17
2.10	The control of photosynthesis	18
2.11	Photosynthesis and the environment	19
2.12	Plants and minerals	20

2d Animal nutrition and health

2.13	Food and ideal diet 1: carbohydrates, lipids, and proteins	21
2.14	Food and ideal diet 2: vitamins, minerals, water, and fibre	22
2.15	Food as fuel	23
2.16	Malnutrition	24
2.17	Animal nutrition	25
2.18	Ingestion	26
2.19	Digestion	27
2.20	Absorption and assimilation	28

2e Circulation

2.21	Water and mineral uptake	29
2.22	Transport systems in plants	30
2.23	Transpiration	31
2.24	The leaf and water loss	32
2.25	Transport systems in animals	33
2.26	The circulatory system	34
2.27	Capillaries: exchange of materials	35
2.28	The heart	36
2.29	Coronary heart disease	37

2f Health and disease

2.30	Disease	38
2.31	Pathogens	39
2.32	Preventing disease	40
2.33	Individuals and the community	41
2.34	Fighting infection: blood and defence against disease	42
2.35	Antibodies and the immune response	43

2g The respiratory system

2.36	Respiration	44
2.37	Contraction of muscles in respiration	45
2.38	The measurement of respiration	46
2.39	Gas exchange	47
2.40	Breathing ventilates the lungs	48
2.41	Smoking and disease (1)	49
2.42	Smoking and disease (2)	50

2h Excretion and homeostasis

2.43	Kidney function	51
2.44	Dialysis	52
2.45	Homeostasis	53
2.46	Controlling body temperature	54

2i Receptors and senses

2.47	Coordination: the nervous system	55
2.48	Neurones in reflex arcs	56
2.49	The central nervous system	57
2.50	The eye	58

2j Hormones, drugs, and tropisms

2.51	The endocrine system	59
2.52	Drugs and disorders of the nervous system	60
2.53	Sensitivity and movement in plants: tropisms	61

3a Plant reproduction

3.1	Sexual and asexual reproduction	62
3.2	Reproduction in flowering plants	63
3.3	Pollination	64
3.4	Formation of seed and fruit	65
3.5	Conditions for germination	66

3b Human reproduction

3.6	Reproduction in humans	67
3.7	The menstrual cycle	68
3.8	Fertilisation	69
3.9	Contraception	70
3.10	Placenta	71
3.11	Feeding babies	72
3.12	Birth	73
3.13	Sexually transmitted infections	74

Contents

3c Inheritance

3.14	Variation and inheritance	75
3.15	DNA and characteristics	76
3.16	DNA and how the genetic code is carried	77
3.17	Cell division	78
3.18	Inheritance	79
3.19	Patterns of inheritance	80
3.20	Inherited medical conditions	81
3.21	X and Y chromosomes	82

3d Variation and selection

3.22	Variation	83
3.23	Causes of variation	84
3.24	Selection	85
3.25	Natural selection	86
3.26	Artificial selection	87

4a Ecosystems, decay, and cycles

4.1	Ecology and ecosystems	88
4.2	Ecology and the environment	89
4.3	Energy flow	90
4.4	Decay	91
4.5	The carbon cycle	92
4.6	The nitrogen cycle	93
4.7	The water cycle	94

4b Populations

4.8	The size of populations	95
4.9	Human population changes	96

4c Microorganisms

4.10	Bacteria in biotechnology	97
4.11	Biotechnology: the production of penicillin	98
4.12	Biotechnology: the production of penicillin	99
4.13	Yeast	100

4d Human impact on ecosystems

4.14	Genetic engineering	101
4.15	Humans and agriculture	102
4.16	Land use for agriculture	103
4.17	Malnutrition and famine	104
4.18	Pollution	105
4.19	Eutrophication	106
4.20	Conservation of species	107
4.21	Science and the fishing industry	108
4.22	Worldwide conservation	109
4.23	Sewage treatment	110
4.24	Saving fossil fuels	111
4.25	Paper recycling	112

5 Practical biology

5.1	Making a model of DNA	113
5.2	Drawing skills: the structure of flowers	115
5.3	Germination	117
5.4	Transpiration experiment	118
5.5	Variables	119

6 Mathematics for biology

6.1	Measurement and magnification	120
6.2	Multiple births	121
6.3	Enzyme experiments	122
6.4	The control of photosynthesis	123
6.5	Ingestion	124
6.6	The leaf and water loss	125
6.7	Individuals and the community	126
6.8	The measurement of respiration	127
6.9	Formation of seed and fruit	128
6.10	Variation and selection	129
6.11	The size of populations	130
6.12	Human population changes	131

7 Revision

7.1	Analysing command words	132

8 Exam style questions

8.1	Exam-style questions	138

9 Project ideas

9.1	Modelling neurones	149
9.2	Modelling the spinal cord	150
9.3	Naming the parts of the body	151

Glossary	153
Answers	155
Data sheets	163

Characteristics and classification of living organisms

1.1 Life and living organisms

1. The seven characteristics of living organisms are **respiration, growth, sensitivity, nutrition, excretion, movement,** and **reproduction**.

   Complete this table by choosing words from this list and writing them opposite their correct meanings.

	Meaning	Characteristic
A	The ability to detect stimuli and make appropriate responses	
B	A set of processes that makes more of the same kind of organism	
C	Removal from an organism of toxic materials, the waste products of metabolism, or substances in excess of requirements	
D	A set of chemical reactions that breaks down nutrients to release energy in living cells	

   [4]

2. To biologists, classification means:

   A giving organisms a name

   B identifying organisms

   C putting organisms into groups

   D describing organisms

   Underline your answer. [1]

3. The following is a list of groups that biologists use to classify living organisms.

   class   family   genus   kingdom   order   phylum   species

   Rewrite the list in the correct hierarchy of classification.

   [3]

Extension

4. Write out the complete hierarchical classification for a human.

5. Scientists in South Africa have recently discovered remains of an organism they have named *Homo naledi*. Suggest what this name tells you about the relationship of this organism to a modern day human.

Characteristics and classification of living organisms

1.2 Using a key for garden birds

1. The drawings show four common birds that came to feed in an English garden.

Parus caeruleus   *Parus major*   *Turdus merula*   *Erithacus rubecula*

   a. State which two birds scientists believe are most closely related and explain your answer.

   ....

   .... [2]

   b. i. Complete this table showing the external features of these birds. One example column has been completed.

Species/feature	All feathers the same colour	Dark stripe along length of body	Large pale areas on sides of head
Erithacus rubecula	X		
Parus caeruleus	X		
Parus major	X		
Turdus merula	✓		

   [3]

   ii. Use the information in this table to complete the following key to identify the four birds.

   1. No large pale areas on head   go to 2

      Large pale areas on head   go to 3

   2. All feathers the same colour   *Turdus merula*

      Feathers of different colours   *Erithacus rubecula*

   3. ....

      ....

   [6]

Extension

2. a. Search the internet for images of orang-utan, chimpanzee, ring-tailed lemur, siamang, grass monkey, purple langur, and aye-aye.

   b. Using *only external features* make a key to distinguish between these animals. (Hint: try to begin with a question that divides this group of seven animals into two approximately equal-sized groups.)

   c. These animals are all **primates**. Humans are also primates. Suggest the most important difference between humans and other primates.

   d. Living organisms can also be classified using evidence from DNA. Use the internet to find how much DNA humans have in common with the other seven primates. Suggest which of these animals is most closely related to humans.

Characteristics and classification of living organisms

1.3 The life of plants

1. a. Match up the following parts of a plant with the function performed by each of them.

Part of plant
Stem
Root
Leaves
Flowers
Fruit

Function
Absorb water and mineral ions
Usually help dispersal of seed, a reproductive structure
Hold leaves in the best position
May be attractive to pollinating insects or birds
Trap light energy for photosynthesis

   [5]

   b. Complete the following paragraphs about the lives of plants. Use words from this list – each word may be used once, more than once, or not at all.

   **algae   angiosperms   autotrophic   cellulose   chlorophyll
   chloroplast   dicotyledons   ferns   herbivorous
   monocotyledons   photosynthesis   respiration   starch**

   All plants contain the light-absorbing pigment called .... This means that plants can be .... – they can make their own food molecules from simple inorganic sources by the process of .... All the members of the Plant Kingdom are made of cells surrounded by a cell wall made of ....

   The Plant Kingdom can be divided into four phyla, ..., mosses, ..., and seed plants. Many of the seed plants have the seed enclosed inside a fruit – they are called ...., and exist in two groups .... (which have leaves with parallel veins) and .... (leaves have branched veins).

   [9]

   c. Plants absorb light energy through their leaves.

   Suggest how you could calculate the leaf surface area of a tree close to your school.

   Extension

Characteristics and classification of living organisms

1.4 Invertebrates in woodland

1. These four animals were among a group of organisms collected from leaf litter lying on the floor of a deciduous woodland.

Ant   Earthworm   Centipede   Mite

   a. Complete the table below to compare the four animals.

	Ant	Earthworm	Centipede	Mite
Number of pairs of jointed legs present				
Are antennae present? (**Yes** or **No**)				

[4]

   b. Use this key to place each of the animals in its correct group.

1. Jointed legs present	go to question 2
No jointed legs	*Annelid*
2. More than four pairs of legs	go to question 3
Four pairs of legs or fewer	go to question 4
3. Body in two main parts, legs not all alike	*Crustacean*
Body made up of many similar segments, with legs alike one another	*Myriapod*
4. 3 pairs of legs present	*Insect*
4 pairs of legs present	*Arachnid*

Write your answers in the table below.

Animal	Classification group
Ant	
Earthworm	
Centipede	
Mite	

[5]

   c. Insects are members of the phylum *Arthropoda*. Humans have never been able to completely exterminate any insect species, although they have tried to eliminate some species which are pests.

   Complete this table to list some species that are harmful and some that are beneficial to humans.

	Name of insect	Reason why it is directly harmful to humans
1		
2		

	Name of insect	Reason why it is beneficial to humans
1		
2		

Extension

Characteristics and classification of living organisms

1.5 Features of vertebrates

1. This table compares some features of **chordate** (**vertebrate**) animals.

   a. Define the term chordate (vertebrate).

   .... [1]

   b. Complete this table.

Chordate	Body covering	Constant body temperature	Parental care of young
(fish)		No	No
(frog)	Moist skin		No
(turtle)	Scales	No	
(bird)	Feathers		Yes
(rabbit)			Yes

   [6]

   Extension

   c. Humans are vertebrates.
      State how many vertebrae are found in a human backbone. Draw a picture of a single vertebra from the lower back of a human.

   d. State **two** functions of vertebrae.

6

Cells and organisation   1.6 Cells

1. a. Many organisms are made up of cells, tissues, and organs.

   Phloem and stamens are examples of structures found in plants.

   State the name of the system to which each of them belongs.

   Phloem ....

   Stamens .... [2]

   b. The diagrams below show a human nerve cell and a palisade cell from a leaf.

   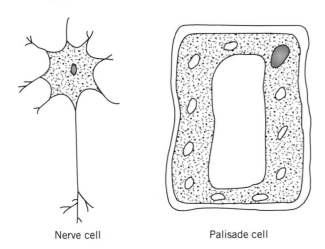

   Nerve cell   Palisade cell

   i. On the diagrams, label **two** features found in both cells. [2]

   ii. State the name of **one** structure found in the palisade cell which allows it to carry out photosynthesis.

   .... [1]

   iii. State the name of **one** other structure found only in plant cells.

   .... [1]

   c. All living cells are able to release energy by the process of respiration.

   i. State the name of the structures in which aerobic respiration takes place.

   .... [1]

   ii. Energy from respiration may be used in protein synthesis.

   State the name of the structures in which protein synthesis takes place.

   .... [1]

   d. i. What is a stem cell?

   ii. Where are stem cells found?

   iii. Use the internet to find why doctors are very interested in stem cells.

Extension

7

Cells and organisation

1.7 The organisation of living organisms

1. The diagrams show several types of plant and animal cells. They are not drawn to the same scale.

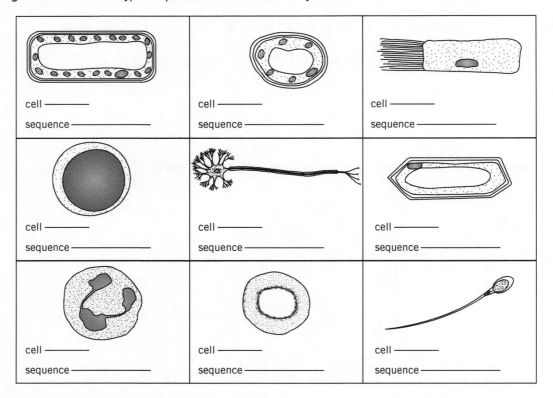

a. Use the key below to identify each of the cells. Write the letter corresponding to each cell on the line next to the appropriate diagram. For each of the cells, write down the sequence of numbers from the key that gave you your answer.

Key

1.	Cell has clear and obvious cell wall	go to 2	5.	Cell has projections at one or more ends	go to 6
	Cell has membrane but no cell wall	go to 4		Cell does not have projections	go to 8
2.	Cell has chloroplasts in the cytoplasm	go to 3	6.	Cell has projections at both ends	CELL E
	Cell does not have chloroplasts in the cytoplasm	CELL A		Cell with projections or projection at only one end	go to 7
3.	Cell with fewer than 10 chloroplasts visible	CELL B	7.	Cell has many projections at one end	CELL F
	Cell with more than 10 chloroplasts visible	CELL C		Cell has a single long projection at one end	CELL G
4.	Cell contains a nucleus	go to 5	8.	Cell has nucleus with many lobes	CELL H
	Cell does not have a nucleus	CELL D		Cell has a round nucleus	CELL I

b. Two of the cells that you have identified would be found in the same plant organ.

State the name of this plant organ. .... [1]

Extension

c. A plant body is made of many specialised cells. All of these cells are made by **cell division** at regions called meristems and then **differentiation** of the cells into different tissues. State where plant meristems are found. Suggest how plant breeders could use meristems in producing many copies of useful plants.

Diffusion and osmosis

2.1 Movement in and out of cells: diffusion

1. a. Complete these paragraphs about the movement of molecules in and out of cells.

Use words from this list. Each word may be used once, more than once, or not at all.

cellulose   diffusion   down   equilibrium   gas   liquid   osmosis
partially permeable   potential   random   rapid   through   up

.... is a process in which molecules move .... a concentration gradient. The movement may take place in a .... or a liquid, and is the result of the .... movement of the molecules. The process continues until an .... is reached.

.... is the .... of water molecules, and takes place down a water .... gradient. This process occurs across a .... membrane.

[9]

> **b.** The water spider carries a bubble of air underwater. The oxygen in the bubble is used up by the spider and replaced with carbon dioxide, but the spider does not have to come up to the surface to replace the oxygen.
>
> From your knowledge of diffusion suggest why the spider does not have to come to the surface to replace the oxygen in the bubble.

Extension

Diffusion and osmosis — 2.2 Movement in and out of cells: osmosis

1. A group of students carried out an investigation into the effects of sugar solutions on rods of potato. The rods of potato were cut using a cork borer, then gently blotted and weighed.

   The rods were placed in groups of three in dishes of distilled water or in one of several sugar solutions of different concentrations. After 4 hours the rods were removed from the solutions, gently blotted, and reweighed.

   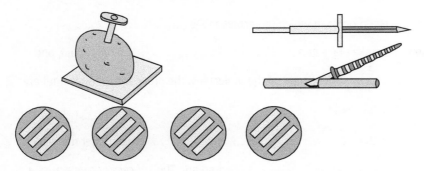

   The students converted the raw results into percentage change in mass for each sugar concentration. The results are shown in the table below.

Concentration of sugar solution (arbitrary units)	Percentage change in mass for rod 1	Percentage change in mass for rod 2	Percentage change in mass for rod 3	Mean percentage change in mass
0 (distilled water)	+8	+8	+8	
0.25	+5	+3	+4	
0.5	0	−2	−1	
1.0	−6	−8	−4	
1.5	−9	−10	−11	

   a. i. Calculate the mean percentage change in mass for each of the different solutions. Write your answers in the right-hand column of the table. [1]

   ii. Plot a line graph of the data on the grid to the right.

   iii. From the graph, calculate the concentration of the sugar solution which would result in no change in mass of the potato tuber.

   .... arbitrary units [1]

   [4]

   iv. Suggest the significance of the value you have obtained for part **iii**.

   ....

   .... [1]

   b. State the name of the process that causes the change in mass of the potato during the course of the investigation.

   .... [1]

   c. Cooks working in kitchens sometimes soak salad vegetables, especially lettuces, in salt water to remove animals such as small flies, caterpillars, and slugs. However, if the lettuce is left in the salt water for too long the leaves become limp and soggy.

   Explain why the lettuce leaves become limp. Suggest how the lettuce leaves could be made crisp again.

Extension

Enzymes and biological molecules

2.3 Organic molecules

1. Complete these paragraphs about some biological molecules.

   Use words from this list. Each word may be used once, more than once, or not at all.

   **amino acids   cellulose   common   DDT   DNA   fatty acids   glucose   glycerol
   glycogen   haemoglobin   insoluble   simple sugars   soluble   sucrose**

   a. Starch consists of smaller units called .... Another molecule made up of these subunits is ...., found in plant cell walls. The sugar most often used for sweetening foods is .... which is so useful because it is ....

   b. Fats are made of smaller units called .... linked to one another by .... Fats are often used for energy storage or as barriers between watery environments – a property that is useful is that fats are .... in water.

   c. Proteins such as .... are made of subunits called .... These subunits are .... so that they are easily transported from one part of the body to another.

   d. Nucleic acids such as .... are made of smaller units that act as a code for protein synthesis in cells.

   [4]

2. Complete this table. Place a + if the element is present and a − if it is not.

	Carbon	Hydrogen	Oxygen	Calcium	Phosphorus	Nitrogen
Protein						
Fat						
Carbohydrate						
Nucleic acid						

   [4]

Extension

3. Find out which is the most common element in the human body.

4. How much calcium is there in 1l of whole milk? How many litres of whole milk would be needed to supply all of the calcium in the body of a 70 kg man?

11

Enzymes and biological molecules

2.4 Testing for biochemicals

1. Soya is a vegetarian meat substitute made from soya beans.

   The table below gives information about the food values of 100 g of soya and 100 g of beef.

Food	Energy content (kJ)	Fat content (g)	Protein content (g)	Carbohydrate content (g)
Soya	1800	23.5	40.0	12.5
Beef	1625	21.0	15.0	6.5

   a. i. Use the information in the table to suggest the main advantage of soya compared with beef.

   .... [1]

   ii. Soya is cheaper to produce than beef.

   Suggest two reasons why soya is cheaper than beef.

   ....

   .... [2]

   b. The nutrient content of different foods can be investigated with a series of simple chemical tests.

   A group of students was given samples of four different powdered foods. They were also given a sample of pure table salt, which only contains sodium chloride.

   They carried out three tests, for glucose, starch, and protein. The table shows the colours produced as a result of their tests.

Food test	Sample W	Sample X	Sample Y	Sample Z	Table salt
Glucose	Red/orange	Blue	Red/orange	Blue	Blue
Starch	Brown	Blue/black	Brown	Brown	Brown
Protein	Purple	Purple	Blue	Purple	Blue

   i. State which one of the samples contained protein but no starch or glucose. .... [1]

   ii. Suggest why table salt was included in the testing. ....

   .... [1]

   iii. Potato contains starch and protein but not glucose. State which food might have been potato.

   .... [1]

Extension

   c. Some popular soft drinks advertise that they contain vitamin C. How could you test a sample of one of these soft drinks to check whether this claim is true? How could you compare the vitamin C content of five different soft drinks?

Enzymes and biological molecules

2.5 Enzymes

1. The two lists show some words referring to enzymes, and definitions of these words.

   Draw lines to match up the terms with their definitions.

Word
Protein
Substrate
Product
Active site
Denaturation
Optimum

Definition
A molecule that reacts in an enzyme-catalysed reaction
The part of the enzyme where substrate molecules can bind
A change in shape of an enzyme so that its active site cannot bind to the substrate
The ideal value of a factor, such as temperature, for an enzyme to work
The type of molecule that makes up an enzyme
The molecule made in an enzyme-catalysed reaction

   [6]

2. This table lists some important uses of enzymes.

Name of enzyme	Use of enzyme
	Part of biological washing powders – removes fatty stains
Restriction enzyme	
Lactase	
	Softens some parts of leather in the clothing industry
	Could break down tough plant cell walls
	Breaks down starch during germination of seeds

   Complete the table by filling in the gaps, using words or phrases from this list.

   amylase   **cellulase**   **clears pieces of tissue from fruit juices**
   cuts out useful genes from chromosomes   **lipase**   **maltase**   **pectinase**
   protease   **releases carbon dioxide during respiration**   **removes milk sugar from milk**

   [6]

   Extension

   3. Potatoes turn brown when they are cut. This browning may be the result of the action of an enzyme on a substance called catechol. Catechol is colourless, but turns brown when it is exposed to air. Browning makes potatoes less attractive to consumers. How might you prevent cut potatoes from turning brown and unattractive? Do other fruits and vegetables also turn brown for the same reason? How could you tell?

Enzymes and biological molecules

2.6 Enzyme experiments

1. Rennin is an enzyme which causes milk to clot. It is used in cheese-making to start making the solid curd.

   A student decided to carry out an experiment on the effect of temperature on the clotting of milk by rennin. She wrote down her results on a scrap of paper.

   <u>Temp and clotting of milk</u>
   60 – nothing
   30°c – only 8 mins.
   15 – nothing again (no clot)
   55 – 7 min
   45 – fastest yet! – 2 min.
   35°c – 5 minute
   20 – 35 mins. to clot
   40 – fast – 3 minutes!
   25 – 18 min.
   50 – 5 min. (same as 35°c!)

   a. Present these results in the form of a suitable table. Draw your table in this space or in your note book. [5]

   b. Plot a graph of these results. Use the grid below.

   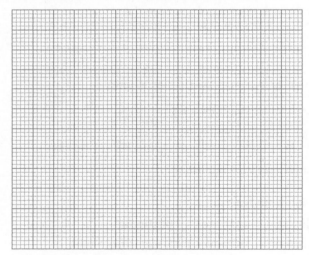

   c. State the optimum temperature for the clotting process ....°C. [1]

   d. i. State one other factor which could affect the activity of the enzyme.

   .... [1]

   ii. Suggest how a student could control this other factor.

   .... [1]

> **Extension**
>
> 2. Find out which enzymes are contained in biological washing powders. Suggest how manufacturers could stabilise enzymes so that the temperatures in the washing machines do not denature them.

Photosynthesis and plant nutrition

2.7 Photosynthesis

1. Complete the following paragraph about plant nutrition.

   The production of food by plants is called .... This process uses

   .... energy trapped by .... in the

   leaves. The process also uses two raw materials from the environment – ....

   from the air and .... from the soil. The first product that can be easily

   detected is .... , a storage carbohydrate. Green plants also produce the

   gas .... during this process – it is released through tiny pores called

   .... [8]

2. The diagram shows the movement of materials in and out of a leaf during photosynthesis.

   materials in and out of leaf

   a. Name the gas entering at **A** .... [1]

   b. Name the gas entering the atmosphere at **B** .... [1]

   c. Name the raw material, required for photosynthesis, entering at **C**.

      .... [1]

   d. Name the mineral, required to manufacture proteins, entering at **C**.

      .... [1]

   e. State the name of the mineral required for the production of the green pigment in the leaves.

      .... [1]

   Extension

   f. It is not easy to measure starch levels in crop plants out in fields. Suggest how a researcher could use a radioactive compound to measure the rate of photosynthesis in crop plants.

15

Photosynthesis and plant nutrition

2.8 The rate of photosynthesis

1. The diagram below shows a method of measuring the rate of photosynthesis of a water plant such as *Elodea*.

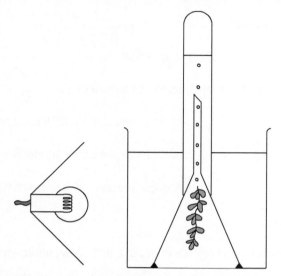

The shoot was exposed to different light intensities and the rate of photosynthesis estimated by counting the number of bubbles released in a fixed amount of time.

The results are shown in this table:

Light intensity (arbitrary units)	1	2	3	4	5	6	7	8
Number of bubbles per minute	6	12	19	25	29	31	31	31

a. Suggest the light intensity at which the plant would have produced 22 bubbles per minute.

.... arbitrary units [1]

b. Explain how the results illustrate the concept of **limiting factors** in photosynthesis.

....

....

.... [2]

Extension

c. A common method for measuring the rate of photosynthesis of an aquatic plant is to collect the volume of gas evolved by an illuminated plant. The method assumes that the gas collected has been released during photosynthesis. State the name of this gas. Suggest how you could test that this gas is the one that you predict it is.

16

Photosynthesis and plant nutrition

2.9 Leaf structure and photosynthesis

1. The diagram shows a section through part of a leaf.

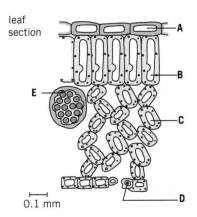

a. Name the cells labelled **A**, **B**, **C**, **D**, and **E**.

Letter	Cell type
A	
B	
C	
D	
E	

[5]

b. State the letter that identifies the cells where most photosynthesis occurs. .... [1]

c. State the letter that identifies the cells which transport water and minerals into the leaf. .... [1]

d. State the name of the carbohydrate that is transported around the plant.

.... [1]

e. Look back at the leaf section. Use the scale to calculate:

i. the thickness of the leaf

ii. the length of a palisade cell.

In each case, show your working. [4]

Extension

2. What is meant by the term leaf mosaic? Explain why this is an important term in understanding the efficiency of a plant in photosynthesis.

Photosynthesis and plant nutrition
2.10 The control of photosynthesis

1. A group of scientists working in an experimental plant research station were interested in how different factors play a part in the control of photosynthesis. They made a series of measurements under different conditions. Their results are shown in the table below.

	Temperature (°C)	Light intensity (arbitrary units)	Carbon dioxide concentration (%)	Rate of photosynthesis (arbitrary units)
A	22	5	0.04	80
B	22	10	0.04	80
C	22	5	0.15	170
D	22	10	0.15	200
E	30	5	0.04	80
F	30	10	0.04	80
G	30	5	0.15	185
H	30	10	0.15	300

a. Describe the set of conditions which gave the highest rate of photosynthesis.

.... [1]

b. The normal concentration of carbon dioxide in the atmosphere is about 0.04%. Use the results in the table to describe the benefits of adding carbon dioxide to the air.

....

.... [2]

c. Explain why the rate of photosynthesis is the same in conditions **A, B, E,** and **F**.

....

.... [2]

> **Extension**
>
> 2. Many types of flowers, fruit, and vegetables are available in supermarkets all year round. This is because they can be grown cheaply in some countries and then flown to the area where they will be consumed. Suggest reasons why importing products like this is a good idea, and reasons why it is potentially harmful to the environment.

Photosynthesis and plant nutrition

2.11 Photosynthesis and the environment

1. Hydrogencarbonate indicator can be used to show changes in the concentration of carbon dioxide in the air.

   CO_2 concentration decreases ← → CO_2 concentration increases

   Purple ← Red → Orange/yellow

   Celandine is a plant that can survive at low light intensities, such as those found on the ground in a forest. Ox-eye daisy cannot survive in these conditions.

   A set of experiments was carried out to investigate the reason for this difference.

   A set of boxes was set up, each lit by a light of a different light intensity. Three glass tubes, as shown in the diagram (right), were placed inside each box. The hydrogencarbonate indicator solution was red in each of the tubes at the start of the experiments.

   After one hour the tubes were removed from the boxes, and the colour of the hydrogencarbonate solution was noted. The results are shown in the table below.

   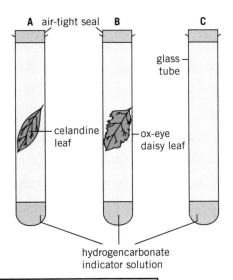

Plant species	Light intensity (arbitrary units)					
	1	4	8	16	64	128
Celandine	yellow	red	red	purple	purple	purple
Ox-eye daisy	yellow	yellow	yellow	yellow	red	purple

   a. Explain why tube **C** is included at each of the light intensities.

   ....

   .... [2]

   b. Exactly the same volume of hydrogencarbonate indicator solution was added to each of the three tubes. Explain the reason for this.

   .... [1]

   c. State which process is occurring most rapidly when the indicator solution:

      i. turns from red to purple .... [1]

      ii. turns from purple to red .... [1]

      iii. turns from red to yellow. .... [1]

   d. Suggest why ox-eye daisies do not occur in forest environments.

   .... [1]

Extension

2. Doctors sometimes advise that houseplants should be provided in hospital wards. Suggest how this could possibly benefit the patients in these wards.

Photosynthesis and plant nutrition

2.12 Plants and minerals

1. The diagrams below show the results of an investigation into the mineral requirements of wheat seedlings.

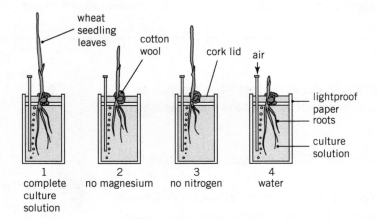

Tube number	Mineral content	Total length of leaves (mm)
1	Complete	24.5
2	No magnesium	11.5
3	No nitrogen	20.5
4	No minerals	5.5

a i. Explain why tubes 1 and 4 were included in the investigation.

   ....

   ....

   .... [2]

   ii. Explain why it is necessary to bubble air through the tubes.

   ....

   ....

   .... [2]

   iii. Explain the results for tube number 2.

   ....

   ....

   .... [2]

b. Farmers often supply these minerals as fertilisers. The bags of fertiliser often have the letters NPK stamped on them.

   State what the initials NPK stand for: .... [1]

> **Extension**
>
> 2. For many years, humans have eaten seaweed as a source of one particular mineral. Which mineral is this, and why is it so important to humans?

Animal nutrition and health

2.13 Food and ideal diet 1: carbohydrates, lipids, and proteins

1. a. This pie chart shows the proportion of different food molecules in a diet.

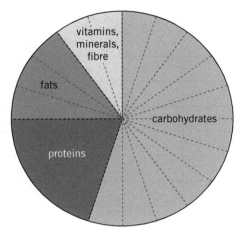

Calculate the percentage of the diet made up of carbohydrates and fats together. Show your working.

[2]

b. Choose whether each of the following statements about food and feeding is TRUE or FALSE.

Statement	True or False
Starch is a carbohydrate found in bread and pasta	
Meat and fish are essential in the diet, as they are the only sources of protein	
Biuret reagent gives a light purple colour with protein	
Oranges and lemons contain vitamin C, and so may help to prevent rickets	
Fats dissolved in alcohol give a milky white colour when mixed with water	
Bacteria can be engineered to provide proteins useful to humans	
Strawberries are a good source of iron	
Vitamin C can be detected using a blue solution called DCPIP	
Assimilation is the process in which foods are transferred from the gut to the blood	
Starch gives an orange colour with iodine solution	
Calcium helps build strong teeth and bones	
Kwashiorkor is a form of malnutrition caused by a low-iron diet	
Too much fat in the diet causes scurvy	
The most common molecule in the human body is water	
Humans cannot eat fungi	

[15]

Extension

c. Protein, a vital part of our diet, is made up of amino acids. How many different amino acids are there? Some of these amino acids are called essential amino acids. What does this mean, and how many are essential for humans? Check why guinea pigs are often used in experiments on human diet.

Animal nutrition and health

2.14 Food and ideal diet 2: vitamins, minerals, water, and fibre

1. a. The table below shows the daily energy requirements of different people.

Person	Energy requirement (kJ per day × 1000)
Eight-year-old boy or girl	4
Teenage girl	12
Teenage boy	14
Manual worker (builder)	18.5
Patient in hospital	7
Male IT worker	11.5
Female IT worker	9.5
Elderly person (aged 75)	8

   i. Use data from the table to suggest **two** factors which influence an individual's daily energy requirements.

   ....

   .... [2]

   ii. Suggest why a female IT worker might have a lower energy requirement than a male IT worker.

   .... [1]

   iii. Elderly people often reduce their food intake as they need less energy. Sometimes this reduction in food intake means that they do not take in sufficient quantities of vitamins and minerals.

   Suggest the likely problems an elderly person might face if the diet were deficient in:

   Vitamin D ....

   .... [2]

   Iron ....

   .... [2]

> **Extension**
>
> b. A balanced diet can be illustrated in the form of a food pyramid. Search for another way of showing the components of a balanced diet as an image. Which do you think is better – the food pyramid or the alternative that you have found?

Animal nutrition and health

2.15 Food as fuel

1. a. State the two components of a balanced diet that provide the most energy.

.... [2]

b. i. Two students, Jack and Sara, each went to the cafeteria to have a meal.

The different meals they chose are shown below.

The total energy content of Sara's meal has been calculated.

Do you think that the energy content of Jack's meal is greater or less than that of Sara's meal? Write More or Less in the space below the list of Jack's meal.

Sara's meal	**Jack's meal**
Chicken (150 g)	Beefburger (120 g)
Jacket potato (150 g)	White bread roll (50 g)
Tomato (40 g) Lettuce (10 g)	Tomato ketchup (10 g)
Cucumber (15 g)	Packet of peanuts (50 g)
Salad cream (15 g sachet)	Glass of full cream milk (200 g)
Fruit yoghurt (150 g)	
Total energy content = 2719 kJ	Total energy content ....

ii. Sara needs about 9000 kJ per day, and Jack needs about 11 000 kJ per day.

Calculate the percentage of Sara's energy needs that would be supplied by this meal.

Show your working.

....%

iii. Suggest **three** reasons why Jack and Sara need different amounts of energy per day.

1. ....

2. ....

3. .... [3]

Extension

2. Recent advice given to consumers has suggested that fats in the diet are less dangerous than refined sugars. If the total fat and sugar content supplies all of the energy needs of an individual, why could the sugars be considered more harmful?

23

Animal nutrition and health

2.16 Malnutrition

1. The table shows the percentage of overweight people in different age groups in a population.

Age group	Percentage of overweight people	
	Male	Female
20–24	22	20
25–29	27	22
30–34	35	27
35–39	40	32
40–44	51	39
45–49	65	44
50–54	64	49
55–59	62	52

a. i. State which sex is more likely to be overweight.

    .... [1]

   ii. State which group has the smallest percentage of overweight people.

    .... [2]

b. i. State the form in which large amounts of extra weight are stored in the human body.

    .... [1]

   ii. Suggest **three** medical conditions associated with being overweight.

    .... [3]

c. Being overweight is one example of malnutrition. Define the term *malnutrition*.

    .... [2]

d. Children in less economically developed countries may suffer from marasmus or kwashiorkor. These conditions are together known as **Protein Energy Malnutrition (PEM)**.

   i. Suggest why a shortage of **protein** can be harmful. ....

    .... [2]

   ii. Suggest the likely effects of a shortage of **energy** foods in the diet.

    ....

    .... [2]

e. Recent reports suggest that the UK, the US, and Saudi Arabia will have very high levels of obesity within ten years. Suggest why these countries, in particular, might experience such high levels of obesity.

Animal nutrition and health

2.17 Animal nutrition

1. Study the two lists below. One is a list of structures in the digestive system and the other is a list of functions of these structures.

   Draw guidelines to link each term with its correct definition.

Structure
Salivary glands
Oesophagus
Stomach
Ileum
Pancreas
Gall bladder
Colon
Rectum

Definition
Produces hydrochloric acid and begins digestion of protein
Produces a set of enzymes which pass into the duodenum
Stores bile produced in the liver
Carries a bolus of food from mouth to stomach
Where most of the water is reabsorbed from the contents of the intestine
Produce an alkaline liquid which lubricates food making it easier to swallow
Stores waste food as faeces
Where most digested food is absorbed

   [8]

2. Read this paragraph, which describes what happens to food in the intestines. Use words from this list to fill in the spaces and complete the paragraph. Each word may be used once, more than once, or not at all.

   bile   bolus   bread   butter   canines   capillaries   carbohydrate   egg
   molars   mouth   pancreas   pH   smaller   stomach   surface area

   An egg sandwich contains starch, fat, and protein. The starch is in the ...., most of the

   fat is in the ...., and much of the protein is in the ....

   When the sandwich enters the .... it is cut into smaller pieces, then crushed by

   the .... This increases the .... so that digestive enzymes can act more

   quickly. The enzymes continue the breakdown of food changing .... molecules such as starch into

   .... soluble molecules. These soluble molecules can pass through the wall of the intestine

   into the ....

   [9]

   Extension

   3. Many bacteria live in the large intestine. What do they do there? How can humans build up the populations of these bacteria?

25

Animal nutrition and health
2.18 Ingestion

1. This diagram shows a section through a molar tooth.

    a. i. Name the parts **A**, **B**, and **C**.

    A ....
    B ....
    C .... [3]

    ii. Name **two** structures found in the pulp cavity.

    1 ....

    2 .... [2]

    b. The diagrams below show three stages in tooth decay.

    stage 1 — enamel attacked, no pain
    stage 2 — decay develops, no pain
    stage 3 — decay develops further, *PAIN*

    i. Suggest why it may take several **months** for a cavity to develop in the enamel of a tooth.

    .... [1]

    ii. Suggest why pain is not felt until stage 3.

    ....

    ....

    .... [2]

    c. Fibres hold the tooth into its socket in the gum. These fibres loosen and teeth may fall out as a symptom of the disease scurvy.

    State the name of the vitamin needed to produce these fibres properly.

    .... [1]

    Extension

    d. How long is a typical human canine tooth? How does this compare with the largest tooth of *Tyrannosaurus rex* or *Carcharodon carcharias*?

26

Animal nutrition and health

2.19 Digestion

1. Human salivary amylase can break down starch. In order to investigate the optimum conditions for starch digestion, four test tubes were set up as shown in the diagram below.

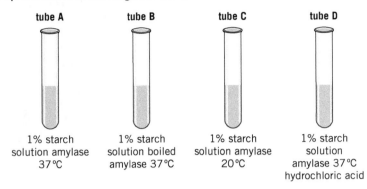

tube A — 1% starch solution amylase 37 °C
tube B — 1% starch solution boiled amylase 37 °C
tube C — 1% starch solution amylase 20 °C
tube D — 1% starch solution amylase 37 °C hydrochloric acid

a. After ten minutes, samples were taken from each of the tubes and tested separately with iodine solution and with Benedict's reagent.

Complete the table below to show the likely results of this investigation.

	Colour in tube A	Colour in tube B	Colour in tube C	Colour in tube D
Tested with iodine solution				
Tested with Benedict's reagent				

[4]

b. State where amylase is secreted in the human digestive system.

.... [2]

c. Explain why tubes **A**, **B**, and **C** were incubated at 37 °C.

.... [1]

d. Suggest where the conditions in tube **D** are likely to occur in the human digestive system.

.... [1]

Extension

2. Humans who suffer from cystic fibrosis often have difficulties with digestion of fats and proteins. Suggest why this is the case. What would be the likely effects of this difficulty on the growth of children with cystic fibrosis?

Animal nutrition and health

2.20 Absorption and assimilation

1. Study the structure shown in the diagram (right).

   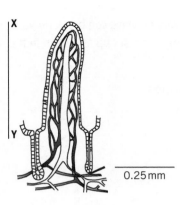

   a. Name this structure. .... [1]

   b. Suggest where this structure would be found.

   .... [1]

   c. i. Draw a line, labelled M, to show the vessel into which fatty acids and glycerol are absorbed. [1]

   ii. Draw a line, labelled N, to show a cell which releases a product to protect the lining of this structure. [1]

   d. Describe how this structure is adapted to its function.

   ....

   ....

   ....

   .... [3]

   e. Use the scale line on the diagram to calculate:

   i. the magnification of this diagram

   [2]

   ii. the actual height of the structure, from **X** to **Y**.

   Show your working in each case. [2]

Extension

2. Some individuals suffer from a condition called Crohn's disease.

   a. Try to find out the cause of the condition.

   b. Explain why one significant symptom of Crohn's disease is a failure to gain weight during growth.

   c. Suggest how this condition can be treated.

Circulation

2.21 Water and mineral uptake

1. These paragraphs refer to transport systems in plants.

   Use words from this list to complete the paragraphs. You may use each word once, more than once, or not at all.

active transport	cambium	diffusion	digestion	
epidermis	hairs	ions	magnesium	nitrate
osmosis	phloem	photosynthesis	respiration	solvent
support	surface area	transpiration	vascular	xylem

   Water is obtained by plants from the soil solution. The water enters by the process of ....

   through special structures on the outside of the root called root .... These structures

   increase the .... of the root. As well as absorbing water they can also take up ....

   such as .... which is required for the production of chlorophyll. These substances are absorbed

   both by .... and by .... (a process that requires the supply of energy).

   Plant cells rely on water for ...., as a .... in chemical reactions, and

   as a raw material for .... Water is also used as a transport medium. [10]

Extension

2. In parts of the world where flooding of low-lying land is a problem, some crops grow very slowly. Explain why poor drainage causes slow growth in crop plants.

Circulation

2.22 Transport systems in plants

1. Study the diagrams below.

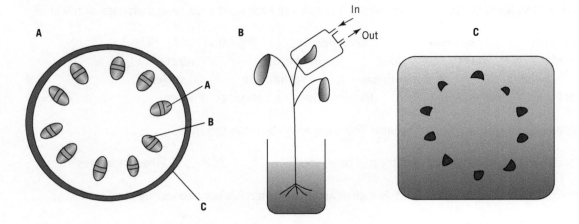

   a. Identify the tissues labelled **A**, **B**, and **C** in diagram A.

   A .... B .... C .... [3]

   b. An experiment was carried out using the plant shown in diagram **B**. The roots of the plant were allowed to stand in a solution of eosin (a red dye), and one of the leaves was kept inside a container that could be filled with radioactively labelled carbon dioxide.

   The apparatus was placed in bright light for 6 hours. A cross-section of the stem was then cut using a sharp scalpel.

   i. State which of the tissues **A**, **B**, or **C** would be stained red. .... [1]

   Explain your answer. ....

   .... [1]

   ii. The section was allowed to stand on a piece of film sensitive to radiation.

   When the film was developed it appeared as shown in diagram **C**.

   Explain why the film had this appearance. ....

   ....

   .... [2]

> **Extension**
>
> 2. Who was the Reverend Stephen Hales? How did he contribute to our knowledge of plant transport systems?

Circulation

2.23 Transpiration

1. The diagram shows the apparatus used by a student in a laboratory investigation of the water balance of a green plant. The roots were carefully washed before the plant was placed in the measuring cylinder.
The apparatus containing the plant was weighed at the start of the investigation and again 24 hours later. The scale on the measuring cylinder was used to read the volume of water. The same investigation was also carried out by four other students.

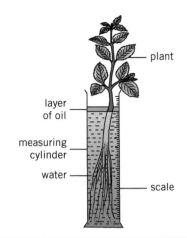

The results of the investigation are shown in the table.

	Mass of apparatus / g					Mean mass / g	Volume of water / cm^3					Mean volume / cm^3
Start	220	225	217	221	219		100	100	100	100	100	
24 h later	208	214	209	210	210		87	89	88	87	89	

a. Calculate the mean loss of mass due to water loss from the plant during the 24-hour period.

Show your working.

.... [3]

b. Calculate the mean **mass** of water which has been absorbed by the roots of the plant during the 24-hour period.
Show your working.

....g [3]

c. Using your knowledge of how water moves upwards in a green plant, explain why your answers to **a.** and **b.** are quite similar.

....
....
.... [2]

d. Explain why the mass of water absorbed by the roots is not exactly the same as the amount of water lost by the plant.

....
.... [2]

Extension

2. Explain why gardeners are advised to water their plants (a) in the evening, and (b) on the soil rather than on the plant's leaves.

Circulation   **2.24 The leaf and water loss**

1. The table lists a series of comparisons between a desert plant and a woodland plant.

	Desert plant	Woodland plant
A	Tissues are more resistant to drought	Tissues are less resistant to drought
B	Leaf system small in comparison to root system	Leaf system large in comparison to root system
C	Stomata often close during daylight	Stomata rarely close during daylight
D	Leaves are often broad and fleshy	Leaves are often thin and rolled
E	Root system has shallow and deep parts	Root system often shallower than desert plant

   a. State which of the comparisons is not correct. .... [1]

   b. Explain your answer. ....

   .... [1]

2. a. The diagram below shows part of a leaf section.

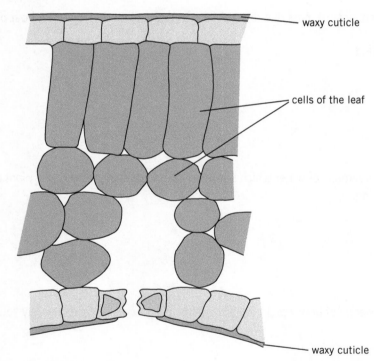

    Annotate the diagram to explain how most of the water is lost from the leaf.   [3]

   b. The leaf was kept at 25°C.

    State **two** factors other than temperature which could affect the rate of water loss from the leaf.

    ....

    .... [2]

3. Why do leaves change colour in autumn and winter?

Circulation

2.25 Transport systems in animals

1. a. The diagram shows the composition of human blood.

   i. Calculate the percentage of plasma in the blood.

   .... [1]

   ii. Name two substances transported in the plasma.

   .... [2]

   iii. Match each of these blood cells with its correct function.

Cell type
Red blood cell
Phagocyte
Lymphocyte
Platelet (cell fragments)

Function
Engulfing invading microbes
Transport of oxygen
Part of clotting process
Antibody production

   [4]

   b. The number of red blood cells tends to increase with altitude. Suggest the symptoms that might be experienced by a climber being flown by helicopter directly, without training, to 3000 m.

   ....

   .... [2]

   c. Athletes who compete in races over distances from 1500 to 10 000 m do better if they live or train at altitudes greater than 2500 m. Suggest a reason for their improved performance.

   .... [1]

   d. Some racing cyclists have used the drug EPO to increase the number of red blood cells they have. State the name of one other drug which has been used to 'cheat' in sport. Explain the biological reason for an athlete using this drug.

   Drug ....

   Reason for use .... [2]

   Extension

   e. There are 5 000 000 red blood cells in each mm^3 of human blood. The total blood volume for an average human is 5 dm^3. All of the red blood cells are replaced every 120 days.

   Calculate how many red blood cells are produced every day. How many are produced every second?

   Each red blood cell contains about 280 000 000 haemoglobin molecules. Calculate how many haemoglobin molecules must be manufactured every second.

Circulation
2.26 The circulatory system

1. The diagram shows a human heart that has been cut across in the region of the ventricles.

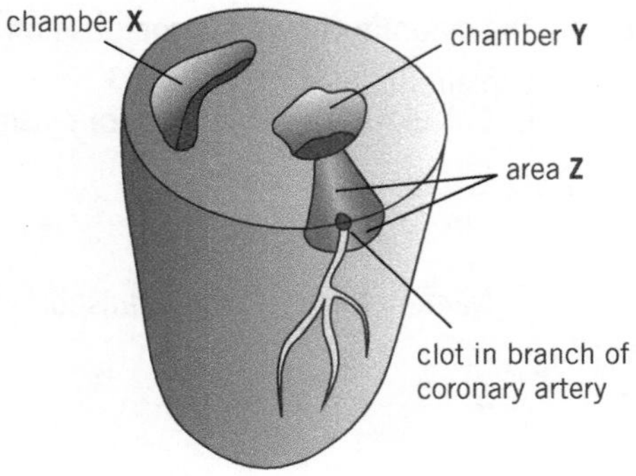

   a. i. State the name of the chamber labelled **Y** .... [1]

      ii. Explain your answer.

      ....

      .... [2]

   b. The beating of the heart is controlled by a patch of tissue called a ....

      In a healthy person the heart normally beats at about .... beats per minute.

2. In mammals, the blood flows through the heart twice for each complete circuit of the body – this is called a

   .... circulation. [1]

3. In contrast, in fish blood flows through the heart only once for each complete circuit.

   The diagram shows the difference between these two types of circulation.

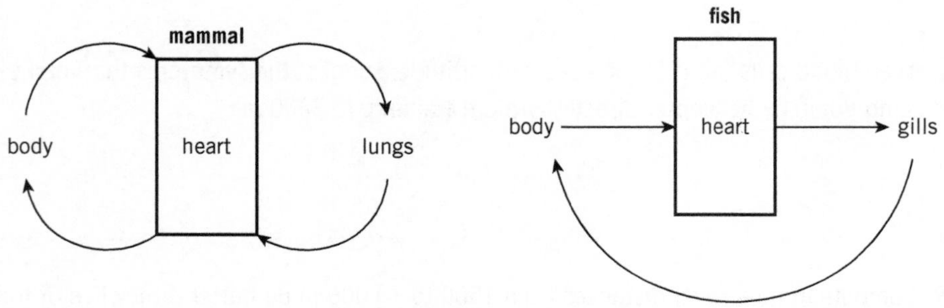

   a. State one **disadvantage** of the single circulation.

      .... [1]

   b. One **advantage** of the double circulation is that pressure is high enough to allow efficient filtration of the blood in the paired .... [1]

   c. The veins of fish tend to be much wider than those of mammals.

      Suggest one reason for this. .... [1]

   d. In a mammal, blood is transported to the lungs in the ....

      .... If pressure in these vessels is too high, ...

      can leak into the lungs. This sometimes happens to climbers at high altitude so the climbers have difficulty in breathing. [2]

Extension

4. Fish have a single circulation, and mammals have a double circulation. What about amphibians?

Circulation
2.27 Capillaries: exchange of materials

1. a. Blood circulation in a mammal is described as a **double circulation**.
   Explain the meaning of the term double circulation.

   ....

   .... [2]

   b. The diagram shows one part of the circulatory system. The blood pressure, in kPa, is marked at three points on the diagram.

   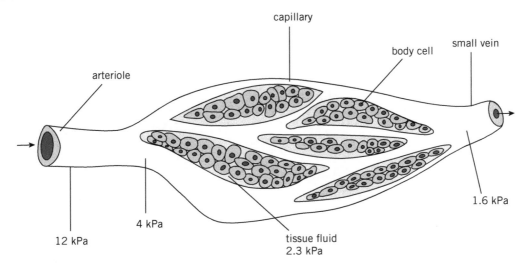

   i. Describe what happens to the blood pressure as blood passes through the arteriole.

   ....

   .... [2]

   ii. State the names of two substances which pass from blood plasma to tissue fluid, and two substances which pass from tissue fluid to blood plasma.

   From blood plasma to tissue fluid .... [2]

   From tissue fluid to blood plasma .... [2]

   c. Suggest what would happen to the blood pressure if a human had a deep cut in the arm.

   .... [1]

   d. Suggest one effect of having a low blood pressure.

   .... [1]

   e. List two features of a capillary that lead to efficient transfer of solutes between blood and tissue fluid.

   ....

   .... [2]

> **Extension**
>
> Antihistamines are medications taken to reduce swelling from insect bites and stings. Explain how they work, using your knowledge of capillaries.

Circulation   2.28 The heart

1. The graph shows the pressure changes in the left side of the human heart, during one complete beat.

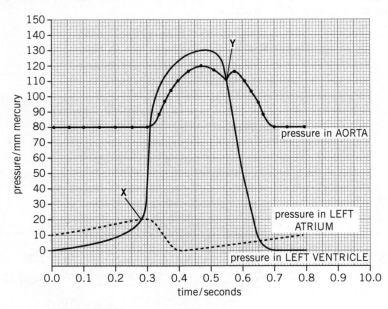

   a. Apart from a change in heart rate, how else can an athlete increase the amount of blood pumped per minute?

   .... [1]

   b. i. From your knowledge of how the heart works, suggest which valves close at points **X** and **Y**. In each case, explain why the valve closes at that point.

   [4]

   ii. Where else in the circulatory system are valves found?

   .... [1]

   iii. What is the purpose of these other valves?

   .... [1]

   c. Suggest why, if the heart is working poorly, a person may have blue lips.

   ....
   .... [2]

   d. One treatment for a failing heart is to have a heart transplant. Following the transplant, the heart may be rejected. Why does this happen?

   e. A transplant patient may be given immunosuppressive drugs. Explain why such a person may be likely to catch simple infections which another person would not suffer from.

   f. Doctors know that there is a shortage of hearts for transplant. They suggest that, in future, people may be given the heart of a pig or a baboon. Why would patients believe that this is not an acceptable procedure?

Circulation

2.29 Coronary heart disease

1. a. A group of students carried out an experiment to investigate their pulse rates.

   Their results are shown in this table.

Name	Data collected	Pulse rate (beats per minute)
Owen	86 beats per minute	86
Sally	231 beats in 3 minutes	
Ayesha	24 beats in 20 seconds	
Amber	80 beats per minute	80
Stephen	16 beats in 10 seconds	
Ahmed	243 beats in 3 minutes	
Alison	38 beats in 30 seconds	
Manfred	34 beats in 20 seconds	
Kristina	36 beats in 30 seconds	
George	178 beats in 2 minutes	

   i. Complete the third column of the table. [2]

   ii. Describe the pattern you can see in these results. ....
   .... [1]

   b. The table shows the rate of blood flow (cm^3 per minute) to organs of the human body, at rest and during exercise.

Organ	Rate of blood flow (cm^3 per minute)	
	Rest	Exercise
Heart	300	400
Muscles	1000	4500
Gut	1500	1000
Kidneys	1100	900
Skin	450	1500
Brain	800	800
Total output	5150	9100

   i. State the effect of exercise on the flow of the blood to the organs of the body.

   ....
   .... [3]

   ii. Explain why the change in blood flow to the muscles is important.

   ....
   .... [2]

   c. Regular exercise reduces the risk of heart disease.

   i. Explain why a blockage in the coronary artery can lead to a heart attack.

   ....
   .... [2]

   ii. State **two** other factors which increase the risk of heart disease.

   ....
   .... [2]

   d. Draw a diagram of the human body to show how a stent is inserted into an artery. Explain why this procedure may be necessary.

Extension

Health and disease

2.30 Disease

1 a. State **one** symptom of a person that has food poisoning.

.... [1]

b. Describe how a mother could spread food poisoning bacteria to her child.

....

....

.... [3]

c. Suggest **one** way a mother could make sure that she did not spread bacteria to her child.

.... [1]

2. Not all diseases are spread from person to person. These diseases are **non-transmissible diseases**.

Non-transmissible diseases may be caused by a number of factors.

Suggest **one** disease which may be caused by **each** of the following factors.

Deficiency in diet .... [1]

Inheritance .... [1]

Age-related degeneration .... [1]

Lifestyle .... [1]

> **Extension**
>
> **3.** What is the Zika virus? Which particular disease is linked to this organism? Suggest how the Zika virus could be controlled.

Health and disease — 2.31 Pathogens

1. **a.** Define the term *pathogen*.

   ....

   .... [2]

   b. Pathogens may be spread in a number of ways.

   Use straight lines to match up pathogens with their methods of transmission.

Pathogen
Cholera bacterium
Influenza virus
Athlete's foot fungus
Plasmodium protoctist
Salmonella bacterium
Human immunodeficiency virus

Method of transmission
By direct contact
In contaminated food
In infected water
In infected body fluids
By an insect vector
In droplets in the air

   [5]

2. This table provides information about some diseases in humans.

Disease	Type of pathogen which causes the disease	How disease is spread	One method of control
AIDS	Virus		Antiviral drugs
Athlete's foot		Spores on damp floors in changing rooms	
Cholera	Bacterium		Provision of clean water supplies
Dysentery		In infected faeces	
Food poisoning		Eating infected food	
Influenza		Droplet infection	

   Complete this table. [10]

Extension

3. Find examples of diseases caused by pathogens using mosquitoes as a vector. Suggest how humans could try to prevent one of these diseases.

Health and disease

2.32 Preventing disease

1. The pie chart shows the number of food poisoning incidents reported from different food sources.

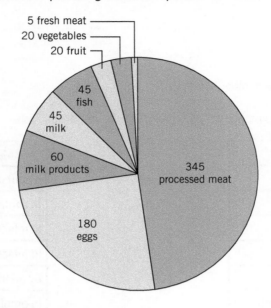

a. Complete this table using the information in the pie chart.

Food source	Number of reported outbreaks
Eggs	180
	45
Fruit	
Milk	
	60
Processed meat	345
	20

[3]

b. Processed meat products include chicken, ham, and corned beef. Packaging on these products usually includes the instruction 'keep refrigerated'.

Explain why refrigeration makes processed food less likely to cause food poisoning.

....

.... [2]

Extension

c. What is the safe temperature in a domestic refrigerator? What is the temperature used to produce pasteurised milk? What about UHT milk?

Health and disease

2.33 Individuals and the community

1. One way in which the community takes responsibility for health is in the provision of a health service. A health service may include screening for disease, and treatment of individuals who are unwell.

   About 30% of individuals in Western Europe and the USA will develop some form of cancer during their lifetime.

   a. State what is meant by **cancer**. ....

   .... [2]

   b. The pie chart below shows the percentage of different types of cancer reported in one population of females.

   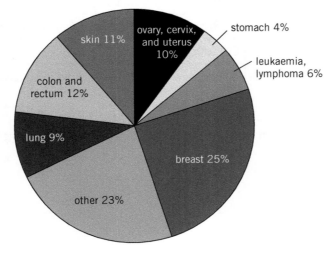

   i. State which is the most common cancer suffered by females in this population.

   .... [1]

   ii. In 2014, one thousand women in this population were reported to have cancer.

   Calculate how many you would expect to suffer from lung cancer.
   Show your working.

   .... [2]

   iii. State one type of cancer, shown in this chart, which would not be reported for males.

   .... [1]

   c. Explain why it is believed that screening for cancer reduces the risk of death from this disease.

   ....

   .... [2]

   d. **Extension** — Scientists believe that monoclonal antibodies may be a possible treatment for some forms of cancer. What is a monoclonal antibody, and how might this type of molecule be useful in cancer treatment?

Health and disease

2.34 Fighting infection: blood and defence against disease

1. Match the following words to their definitions.

Word
Pathogen
Transmissible disease
Antigen
Skin and nasal hairs
Active immunity
Passive immunity
Phagocyte
Lymphocyte

Definition
External barriers to infection
Defence against a pathogen by antibody production in the body
White blood cell that engulfs pathogens
Short-term defence by provision of antibodies from another individual
A disease-causing organism
A substance which triggers the immune response
Cell which produces antibodies
Condition in which the pathogen can be passed from one host to another

[8]

2. This graph shows the level of antibody in the blood in response to two doses of a vaccine.

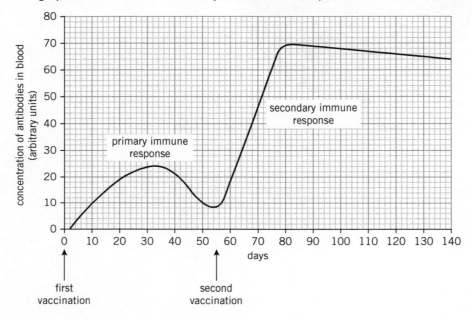

a. Define the term **vaccine**. ....

....

.... [2]

b. State **two** ways in which the response to the second injection is different to the response to the first injection.

....

.... [2]

Extension

3. Some human patients are prescribed the drug **warfarin**. What does this drug do? Which dangerous side effect might it have? What was the original use of warfarin?

Health and disease

2.35 Antibodies and the immune response

1. A vaccine has been developed against the organism that causes TB (tuberculosis). The vaccine has been available for use in many parts of Africa since the late 20th century.

   a. Vaccination against TB is an example of **active immunity**.

      Define the term active immunity. ....

      ....

      .... [2]

   b. State **three** ways in which **passive immunity** differs from active immunity.

      ....

      ....

      .... [3]

   Complete the table below to show whether each method is active or passive.

Method of gaining immunity	Type of immunity (active or passive)
Vaccination against meningitis	
Antibodies cross the placenta and enter the fetus	
A person bitten by a dog is given a serum	
A baby feeds on colostrum (first milk)	
A person becomes infected with a pathogen	

Extension

2. Type 1 diabetes may be an **autoimmune disease**. What is meant by autoimmunity? Find the name of another important autoimmune disease.

   How can the immune response be harmful to an unborn baby?

   ....

The respiratory system 2.36 Respiration

1. Use the clues below to complete the crossword.

Across:

5 Together with 16 across - a 'packet' of energy - ATP for short
6 A form of energy which may be 'lost'
7 One use of energy which allows organisms to get bigger
9 A method of transport which can move molecules against a concentration gradient
10 Could be defined as 'release of energy from foodstuffs'
12 Organelles in which 10 across occurs most efficiently
13 A chemical process which often requires molecular oxygen
14 Involves replication and separation of chromosomes
15 A change in position of an organism or part of an organism
16 See 5 across
18 The most common substrate for 10 across
19 With 17 down, a metabolic process which joins together amino acids with the use of energy
20 In the absence of oxygen
22 In the presence of oxygen
23 The main source of energy on the Earth
24 The organ in which 10 across can provide the heat to maintain body temperature

Down:

1 Another way of saying 'from the Sun'
2 Muscle..requires a supply of energy
3 Required to carry out the metabolic work in a living organism
4 The most important energy conversion in living things
8 Can be carried out if a supply of energy is available
11 A unit commonly used to describe a quantity of 3 down
17 See 19 across
21 The gas necessary for much of life on Earth – 20% of the atmosphere

Extension

2. Cosmonauts return from the Moon with rock samples, and robot space missions bring material from other planets. Scientists suggest that the first sign of life that should be tested for is respiration. Do you agree? Explain your answer.

The respiratory system — 2.37 Contraction of muscles in respiration

1. During exercise, muscle contraction requires energy. Some of this energy is released during aerobic respiration.

   a. i. Write a chemical equation for aerobic respiration.

   [3]

   ii. Suggest why respiration is affected by temperature.

   .... [1]

   b. If the muscles work very hard, the cells may use up all of the available oxygen. When this happens, energy can still be released by anaerobic respiration. These pie charts show the relative amounts of aerobic and anaerobic respiration during a 100 m race and during a 1500 m race.

   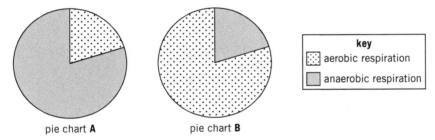

   pie chart **A**   pie chart **B**

   key:
   - aerobic respiration
   - anaerobic respiration

   i. State and explain which of the pie charts shows the results from a 1500 m race.

   ....
   ....
   .... [2]

   ii. State the effect of a build-up of lactic acid on an athlete's performance.

   ....
   .... [2]

   iii. Explain how the level of lactic acid in the muscles is reduced after a race.

   ....
   ....
   .... [2]

   c. Explain what is meant by an **oxygen debt**.

   ....
   ....
   .... [2]

Extension

2. Top athletes can improve their V_{O_2} max. What is this quantity, and why is it so important for endurance athletes?

The respiratory system — 2.38 The measurement of respiration

1. This piece of apparatus is used to study gas exchange.

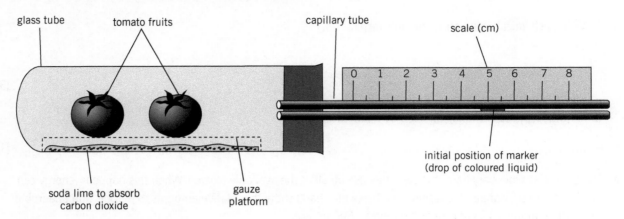

   a. During the course of the experiment the marker moves.

      i. State the name of the process, carried out by the cells of the tomatoes, which causes this movement.

      .... [1]

      ii. State the direction in which the marker will move. .... [1]

      iii. Explain your answer to part **ii**. ....

      .... [1]

      iv. Calculate the position of the centre of the marker after 40 minutes, if the marker moves 0.25 cm every five minutes. Show your working.

      [2]

      v. Suggest a suitable control for this experiment.

      ....

      .... [1]

   b. The class teacher told the students that fruit growers often enclosed the fruits they collect in plastic bags filled with nitrogen gas.

      Explain why this is important.

      ....

      ....

      .... [2]

   c. **Extension** Temperature affects the rate of respiration. Suggest why. How could you check whether your suggestion is a valid one?

46

The respiratory system
2.39 Gas exchange

1. This diagram shows an alveolus (airsac) and a blood capillary.

   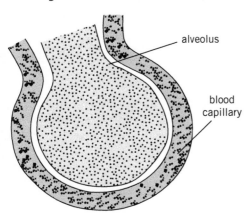

   a. On the diagram, use arrows and suitable labels to describe the movement of oxygen and carbon dioxide during gas exchange. [2]

   b. List **three** features which make alveoli adapted to gas exchange.

      1. ....

      2. ....

      3. .... [3]

   c. State the name of the artery that brings deoxygenated blood to the capillary.

      .... [1]

Extension

2. What is **pleurisy**? How does it affect human health?

47

The respiratory system
2.40 Breathing ventilates the lungs

1. The diagram shows the position and shape of the human thorax (chest) during inhalation (breathing in).

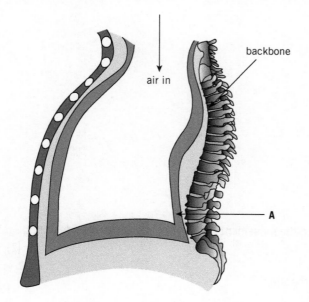

   a. State the name of the part labelled **A**.

   .... [1]

   b. On the diagram label the diaphragm. Use a guideline and the letter **D**. [1]

   c. The diaphragm contains muscles involved in breathing. Name another set of muscles involved in breathing in.

   .... [1]

   d. Describe and explain the how muscles are involved in exhalation (breathing out).

   ....

   ....

   ....

   ....

   .... [3]

> **Extension**
>
> 2. What is the likely vital capacity of a 16-year-old male? Try to find out the vital capacity of Chris Froome, twice winner of the Tour de France.

The respiratory system
2.41 Smoking and disease (1)

1. a. Study this histogram.

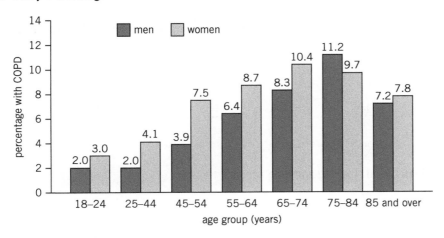

   i. Suggest a suitable title for the chart.

   .... [1]

   ii. State what the initials **COPD** stand for.

   .... [1]

   iii. Does the chart provide evidence that smoking increases the risk of COPD?

   Explain your answer. ....

   ....

   .... [2]

b. The maximum volume of air that can be exchanged with a single breath in and out is called the vital capacity. A typical healthy male would have a vital capacity of about 5000 cm^3.

   Describe and explain the likely effect of long-term smoking on vital capacity. ....

   ....

   ....

   .... [2]

Extension

2. How does nicotine affect the nervous system? Explain how drugs such as nicotine cause physical addiction.

The respiratory system — 2.42 Smoking and disease (2)

1. Cigarette smoke contains carbon monoxide, which can combine with haemoglobin in red blood cells.

   Explain how you would use your biological knowledge and this additional information to persuade a young woman to stop smoking during pregnancy.

   ....

   ....

   .... [2]

2. This bar chart shows a relationship between smoking and lung cancer.

   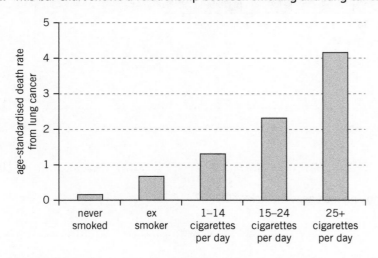

   a. Calculate how much more likely a person who smokes 20 cigarettes per day is to develop lung cancer than a person who has never smoked. Show your working.

   [2]

   b. Some people say that 'just one cigarette a day won't do me any harm'. Use information from the bar chart to help to convince them that this is not the case.

   ....

   ....

   .... [2]

Extension

3. The link between smoking and lung disease has been established by **epidemiology**. Explain another link between a disease and a particular lifestyle that has been established using epidemiology.

Excretion and homeostasis

2.43 Kidney function

1. Three test tubes were set up as shown in the diagram below.

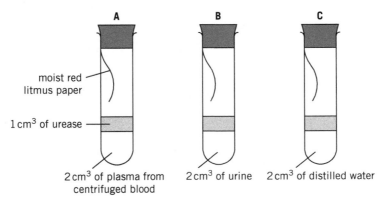

The enzyme urease breaks down urea to release ammonia.

$$\text{urea} \longrightarrow \text{carbon dioxide} + \text{ammonia}$$

The tubes were incubated at 37 °C for 30 minutes. After 30 minutes the contents of the tubes are subjected to several tests. The results are shown below.

	A	B	C
Test substance			
Biuret reagent	Blue to violet	Remains blue	Remains blue
Benedict's reagent	Blue to orange-red on heating	No change on heating	No change on heating
Moist red litmus paper	Turns blue	Turns blue	No change

a. State the name of the substance that the tests show is present in the plasma and in the urine.

.... [1]

b. Explain why the test tubes were incubated at 37 °C.

....

.... [2]

c. i. Identify which substances are found in blood plasma but not in urine or distilled water.

.... [2]

ii. Explain why these substances are not lost from the body.

....

.... [2]

d. State the function of tube **C**.

.... [1]

e. The kidney tubule is called the nephron. How is the length of the nephron different in a gerbil (a desert mammal) from a beaver (an aquatic mammal)? Explain this difference.

Excretion and homeostasis
2.44 Dialysis

1. Dialysis can help people whose kidneys do not function efficiently. Their blood goes through an artificial kidney machine, where it passes alongside a series of partially permeable membranes. On the other side of the membranes is a solution into which waste substances diffuse.

   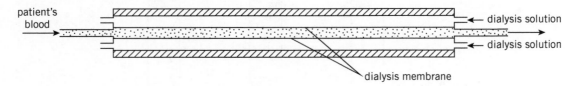

   a. The tube of dialysis membrane is usually coiled rather than straight. Explain why this is important.

   .... [1]

   b. The tube leading out of the machine is usually narrower than the tube leading from the patient's blood into the machine.

      State what effect this will have. ....

      .... [1]

   c. This table compares the concentration of various substances in the blood and the urine of a healthy person.

	Plasma concentration (g per 100 cm³)	Urine concentration (g per 100 cm³)	
Water	910	950	
Protein	75	0	
Glucose	1.0	0	
Urea	0.3	20.5	
Sodium ions	3.1	3.7	
Chloride ions	3.5	6	

   i. State how the composition of the plasma of a patient requiring dialysis would differ from that of a healthy person.

   .... [2]

   ii. Which of the following substances would be able to cross the dialysis membrane? Underline your choice(s).

   ammonia   glucose   protein   red blood cells   urea

   [1]

   iii. Suggest the composition of a suitable dialysis fluid. Write your answers in the shaded column of the table above. You need only to write **higher, lower**, or the **same**. [6]

   d. A person treated for kidney disease by kidney transplantation is usually prescribed immunosuppressive drugs. Explain why these drugs are important, and what side effects they might have.

Excretion and homeostasis

2.45 Homeostasis

1. The diagram shows a section through human skin.

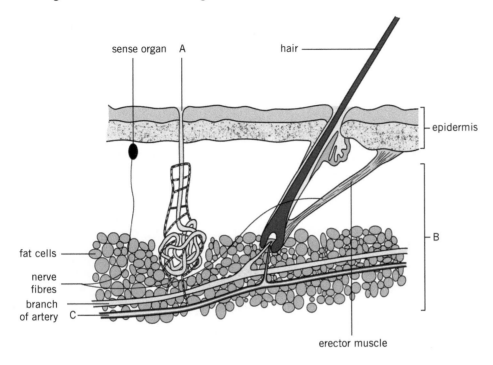

   a. State the name of the parts labelled **A**, **B**, and **C**.

   A .... B .... C .... [3]

   b. State and explain the function of the fat cells.

   ....

   ....

   .... [2]

   c. The body temperature of a mammal remains constant within narrow limits in spite of changes in the environmental temperature.

   State **two** ways in which this 'constant' body temperature is valuable to mammals.

   1. ....

   ....

   2. ....

   .... [2]

   d. Many older Biology textbooks state that a lizard is cold blooded. Is this always the case? Explain your answer.

Excretion and homeostasis

2.46 Controlling body temperature

1. The diagram below shows a section through human skin in warm conditions.

   a. State the names of the structures labelled **A** and **B**.

   A ....

   B .... [2]

   b. i. State what happens to **B** if the erector muscle receives a stimulus to contract.

   .... [1]

   ii. Explain how this response helps to regulate body temperature in a cold environment.

   ....

   ....

   .... [2]

   iii. On the diagram (right) the blood capillary is not complete. Complete the diagram to show the size and position of the capillary in cold conditions.

   [2]

   c. The control of body temperature is an example of **negative feedback**. Explain what is meant by the term negative feedback.

   ....

   ....

   .... [2]

   Extension

   d. One of the symptoms of Type 1 diabetes may be a loss of sensitivity in the feet. Excessive alcohol consumption may damage peripheral nerves and so also reduce sensitivity in the fingertips, toes, and in the small of the back.

   Devise a simple test to check for this altered sensitivity. Make sure that you include some form of control in your proposed test.

Receptors and senses | **2.47 Coordination: the nervous system**

1. This diagram shows a single neurone.

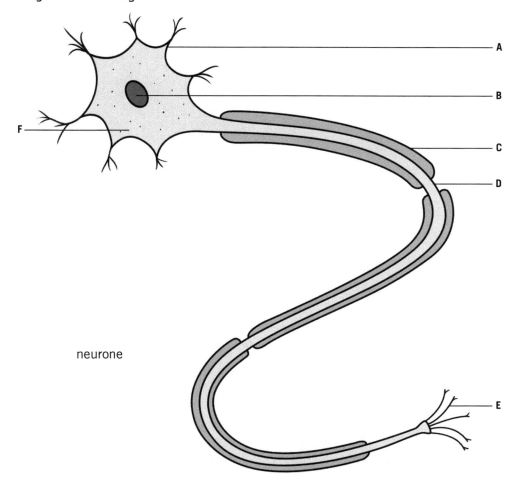

neurone

Match the letters to the descriptions of each of these parts.

Description of part	Letter corresponding to part
Connects with another neurone	
Insulates the neurone to prevent interference with other neurones	
Contains high concentrations of neurotransmitters	
Connects with an effector, such as a muscle	
Allows an impulse to 'jump' quickly along the axon	
Contains DNA	

[5]

Extension

2. Serotonin is a neurotransmitter. Find out what it does, and explain why serotonin-specific reuptake inhibitors are important medicines.

55

Receptors and senses — 2.48 Neurones in reflex arcs

1. a. i. Sneezing is a reflex action. Name the organ containing the receptor cells which detect the stimulus that leads to sneezing.

.... [1]

ii. Suggest **one** advantage of reflex actions.

....

.... [1]

iii. State **two** features of reflex actions.

....

.... [2]

b. i. The diagram below shows a reflex arc.

Select words from this list to name the parts numbered 1–7.

**association neurone   dorsal root   grey matter   motor neurone   receptor
sensory neurone   spinal cord   synapse   ventral root   white matter**

Structure	Name of structure
1	
2	
3	
4	
5	
6	
7	

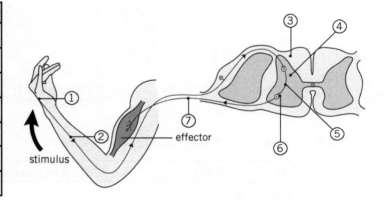

[7]

ii. Muscles and glands are effector organs. State how muscles and glands react when they are stimulated.

Muscles ....

Glands .... [2]

---

c. Are the following statements true or false?

Reflex actions cannot be over-ruled because each of them has a definite survival value.

Some reflex actions have their links in the brain itself.

Reflexes are so important that we don't even know that they have happened.

Receptors and senses
2.49 The central nervous system

1. The apparatus shown below was used by a group of students to check reaction times. The students measured how long it took their fellow students to respond to a red traffic light.

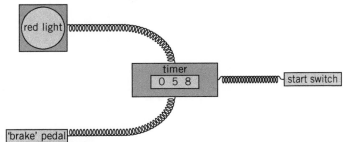

- One student presses the switch.
- The red light comes on.
- A second student has to press the brake pedal as quickly as possible.
- The timer measures the reaction time in hundredths of a second.

The results for the whole class are shown in this table.

Student number	Reaction time (hundredths of a second)	Student number	Reaction time (hundredths of a second)	Student number	Reaction time (hundredths of a second)
1	22	11	29	21	28
2	28	12	23	22	18
3	16	13	24	23	32
4	21	14	19	24	25
5	20	15	34	25	37
6	23	16	18	26	31
7	35	17	27	27	25
8	27	18	19	28	31
9	15	19	23	29	24
10	27	20	26	30	24

a. Complete this table to show the distribution of reaction times in this group of students.

Reaction time (hundredths of a second)	15–19	20–24	25–29	30–34	35–39
Number of students					

b. Calculate i. The mean value for the reaction time. Show your working.

   Mean reaction time .... [2]

   ii. The percentage of students with a reaction time greater than 24/100th of a second.

   ....% [1]

c. If a car was being driven at 60 km per hour, calculate how far the car would travel before the student with the slowest reaction time hit the brake pedal. Show your working.

[3]

Extension

2. Back pain can be very severe. One way of controlling back pain, or the pain during childbirth, is by injecting an anaesthetic under the surface of the spinal cord (an **epidural** injection). How do you think this might help to control pain? (Clue: think about what sensory neurones do.)

57

Receptors and senses — 2.50 The eye

1. The diagram shows a section through the eye.

    a. Use the table below to match up the lettered structures with the descriptions of their functions.

Description	Letter
Contains rods and cones	
Helps to converge light towards the retina	
Is black to prevent internal reflection of light	
Is tough enough to act as an attachment for the muscles that move the eye in its socket	
Is a muscle controlling the amount of light entering the eye	
Contains neurones leading to the visual centre in the brain	

    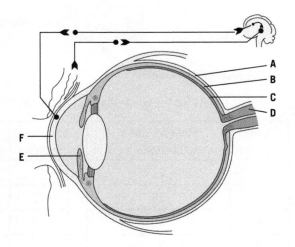

    [6]

    b. The diagram also shows the pathway of an important reflex involved in protection of the eye.

        i. Complete and rearrange the boxes to show that you understand the pathway of this action.

        receptor is ....

        effector is ....   stimulus is ....

        response is ....   coordinator is ....

        [6]

        ii. State the survival value of this reflex.

        .... [1]

2. **[Extension]** What are the effects of ageing on the structure of a human eye? Suggest how these effects can be kept to a minimum.

Hormones, drugs, and tropisms

2.51 The endocrine system

1. a. Define the terms.

   Hormone .... [1]

   Target organ .... [1]

   b. State **two** ways in which control by hormones is different from control by the nervous system.

   1. ....

   2. .... [2]

2. Blood glucose levels increase sharply following a fright or shock.

   a. Explain why this happens. ....

   ....

   ....

   .... [2]

   b. Suggest **two** other changes which could occur in the functioning of the body because of the fright.

   1. ....

   ....

   2. ....

   .... [2]

Extension

3. Athletes in endurance events are often recommended to consume 'slow release' carbohydrates. What are these, and why are the athletes given this advice?

Hormones, drugs, and tropisms

2.52 Drugs and disorders of the nervous system

1. The diagram below shows some of the effects of drinking alcohol.

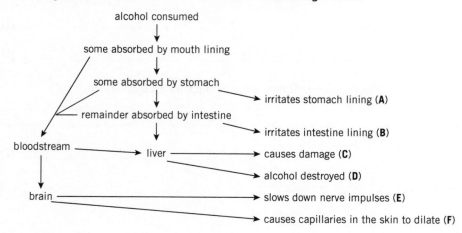

   a. Use letters from the diagram to complete the following table. The table links effects to damage caused to body processes. There may be more than one letter which corresponds to a particular example of damage.

Damage to....	Letter(s) of effect(s)
urea production	
digestion of protein	
absorption of fatty acids	
control of body temperature	

   [4]

   b. Alcohol may affect an unborn child.

   Explain how alcohol may reach the cells of a fetus. ....

   ....

   .... [3]

2. Complete the following table about some drugs. Use phrases from the following list.

Name of drug	Effect on the human body
Heroin	
Nicotine	
Oestrogen	
Penicillin	
Testosterone	

   Protects against syphilis

   Gives a feeling of calmness and rest

   Stimulates the heart rate

   Protects against AIDS

   Increases growth rate in muscle

   Stimulates ovulation [6]

Extension

3. Endurance athletes sometimes 'cheat' by using a drug called EPO. What does this drug do, and how does it help the athletes? How can the use of this drug be dangerous to the athletes?

60

Hormones, drugs, and tropisms

2.53 Sensitivity and movement in plants: tropisms

1. Auxin is a hormone made by the tips of plant shoots.

   a. A shoot was grown in a container so that light shone onto it from one side only. This set of diagrams shows the movement of the auxin in the shoot, and the results of the experiment.

   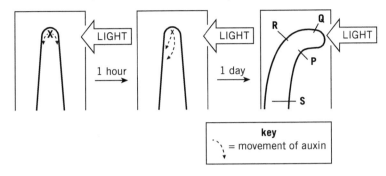

   i. Describe the movement of auxin in the shoot after one hour.

   .... [1]

   ii. Use the diagram to explain how the movement of auxin causes the plant response seen in this investigation.

   ....

   ....

   ....

   .... [2]

   iii. Name the plant response seen above, and explain its value to a plant.

   Name of response .... [1]

   Benefit to plant .... [1]

   b. Plant hormones can be used by farmers to manage plant growth.

   Give **two** examples of the commercial value of plant hormones.

   i. ....

   ....

   ii. ....

   .... [2]

2. If a rug or sheet of corrugated iron is placed on a vegetated patch of land, the growth of the plants will be affected. Suggest how the appearance of the plants would change if the rug or sheet of metal is left lying on the ground for a month.

Plant reproduction — 3.1 Sexual and asexual reproduction

1. The diagrams below show simple schemes for asexual and sexual reproduction.

   asexual

   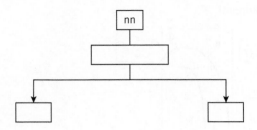

   sexual

   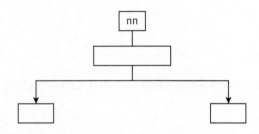

   a. Insert the words **meiosis** and **mitosis** in the correct positions on the diagram. [2]

   b. Write in the correct chromosome number in the boxes on the diagram. [2]

   c. Suggest one **advantage** and one **disadvantage** of asexual reproduction in plants.

   Advantage ....

   Disadvantage .... [2]

   Extension

   d. Explain how both sexual and asexual reproduction could be important in the development of a new crop plant.

2. Scientists have been able to produce clones of laboratory animals such as rats. Suggest how a clone such as this could be useful in the testing of experimental drug treatments.

Plant reproduction — 3.2 Reproduction in flowering plants

1. Most of the variations in flower structure are related to the methods of pollination.

    a. Define the term **pollination**. ....

    .... [1]

    b. This diagram shows a typical wind-pollinated flower.

    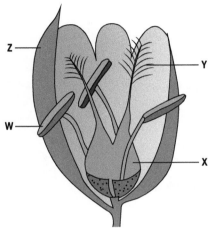

    Identify the structures labelled **W, X, Y,** and **Z**.

    W .... X ....

    Y .... Z .... [4]

    c. Complete this table to compare the structure of wind- and insect-pollinated flowers.

Part of flower	Wind-pollinated	Insect-pollinated	Explanation
Petals			
Anthers			
Pollen			
Stigmas			

    [8]

2. Most plants are **hermaphrodite**, but not all are. Explain why **(a)** not all holly trees bear holly berries, and **(b)** why more than one holly tree must be present in an area for any berries to be formed.

Plant reproduction — 3.3 Pollination

1. The diagram below shows a half of a tomato flower.

   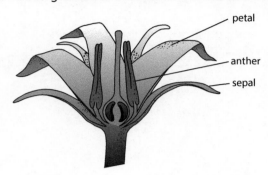

   a. i. Mark with a **P** the part where a pollen grain must land to pollinate the flower. [1]

   ii. State the name of the part where the pollen grain has landed.

   .... [1]

   b. This diagram shows a pollen grain with a pollen tube that is growing towards an ovule.

   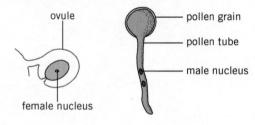

   i. Complete the drawing of the pollen tube to show how it enters the ovule. [2]

   ii. Describe and explain what happens to the male and female nucleus at fertilisation.

   ....

   ....

   ....

   .... [3]

Extension

2. Many fruit growers are anxious about the use of insecticides in areas close to fruit farms.

   Why do some farmers use insecticides? Suggest why the fruit farmers, in particular, are so worried about their overuse.

Plant reproduction   3.4 Formation of seed and fruit

1. The diagram shows a carpel (the female part of a flower).

   a. Match the letters on the diagram with the names of structures in this table.

   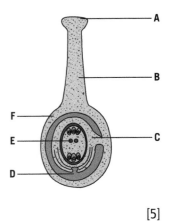

Name of structure	Label letter
Egg cell (female gamete)	
Ovule	
Style	
Stigma	
Micropyle	
Ovary	

   [5]

   b. Complete the diagram to show how a male nucleus fertilises the egg cell. Add labels to your diagram.   [3]

   c. This diagram shows the appearance of a broad bean seed which has been sectioned and stained with iodine solution.

   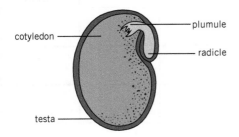

   i. State which parts of the broad bean seed make up the embryo.

   ....   [1]

   ii. State the name of the main food material stored in the broad bean seed.

   .... [1]

   iii. Explain what change you would expect to see in the appearance of the seed to make you draw this conclusion.

   .... [1]

Extension

2. Many biologists are worried about neonicotinoid pesticides. Explain how overuse of these substances could reduce the production of apples and pears in orchards.

Plant reproduction — 3.5 Conditions for germination

1. Six boiling tubes are available (like those shown in the diagram below). Different items can be added to the boiling tubes so that different conditions would exist in each of them.

   The tubes are placed in a bright position.

   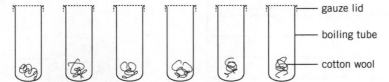

   The tubes were set up as shown in the table below, and left for 48 hours.

   a. You can add to the drawings of the tubes to show what they might look like at the start of the investigation.

Item/condition	Tube A	Tube B	Tube C	Tube D	Tube E	Tube F
Three pea seeds	✓	✓	✓	✓	✓	X
Black paper cover	✓	X	X	X	X	X
Temperature	warm	warm	warm	cool	warm	warm
Pyrogallol added (removes oxygen)	X	✓	X	X	X	X
Dampness of cotton wool	yes	yes	yes	yes	X	yes
Did seeds germinate?						

   Unfortunately you have mixed up your results! Two of the tubes contained germinated seeds, and four did not.

   Write **YES** or **NO** in the boxes to complete the table. [6]

   b. State the conclusions you can make about the conditions necessary for germination.

   ....

   ....

   ....

   ....

   ....

   ....

   .... [4]

Extension

2. The action of enzymes in germinating seeds is affected by temperature. Explain why strict control of temperature is important in the malting stage of the brewing industry.

Human reproduction — 3.6 Reproduction in humans

1. The diagram below shows a mature human sperm.

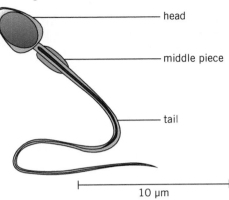

   a. The tip of the head of the sperm contains enzymes that break down proteins. State the function of these enzymes during fertilisation.

   .... [1]

   b. The middle piece of the sperm is packed with mitochondria.

   Suggest a function for these mitochondria. ....

   .... [1]

   c. Use the scale provided to calculate the length of the sperm, from the tip of the head to the end of the tail. Show your working.

   .... [2]

   d. The mass of the nucleus of the sperm is about 3×10^{-6} µg. The nuclei of most other cells in the human body have approximately twice that mass.

   Suggest a reason for this difference. ....

   .... [1]

   e. Female gametes are released at ovulation. The female gamete (ovum) is quite different to the sperm.

   Complete this table to compare sperm and ova.

Feature	Sperm cells	Egg cells (ova)
Site of production		
Numbers produced		
Mobility		
Relative size		

   [4]

Extension

2. Testosterone is necessary for the development of the male reproductive system. Many older men suffer from an enlarged prostate gland. Suggest the likely symptoms of an enlarged prostate, and explain why testosterone inhibitors may reduce these symptoms.

Human reproduction
3.7 The menstrual cycle

1. a. The calendar below shows the menstrual cycle for a woman in June 2016.

Monday	Tuesday	Wednesday	Thursday	Friday	Saturday	Sunday
		1	2	3	4	5
6	7	8	9	10	11	12
13	14	15	16	17	18	19
20	21	22	23	24	25	26
27	28	29	30			

   key: ⊠ = ovulation (the release of an egg)   ▨ = menstruation   ▥ = when the egg is in the oviduct

   i. State how many days of menstruation are shown during June 2016.

   .... [1]

   ii. Assume that this woman has identical menstrual cycles from month to month. State the date in early July 2016 that would be the last day of menstruation.

   .... [1]

   iii. Explain why fertilisation could not occur on the 7th June 2016 even if active sperms are released into the vagina.

   .... [1]

   b. The changes in the thickness of the uterus wall during the menstrual cycle are affected by the hormones progesterone and oestrogen.

   i. State which hormone reaches its maximum 1 to 2 days before menstruation.

   .... [1]

   ii. State which hormone reaches its maximum 1 to 2 days before ovulation.

   .... [1]

   c. The hormone oestrogen is also involved in the changes to the female's body that occur at puberty.

   i. State the name of the equivalent male hormone that influences the male body at puberty.

   .... [1]

   ii. List three changes that occur in the male body at puberty.

   ....

   ....

   .... [3]

Extension

Human reproduction — 3.8 Fertilisation

1. The diagram below shows a front view of the human female reproductive system.

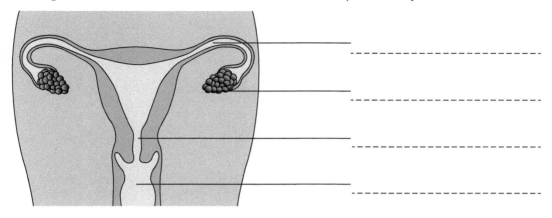

   a. Add labels to the diagram. [4]

   b. i. Place a letter F on the diagram to show where the sperm usually fertilises the egg. [1]

      ii. State the name of the structure formed when an egg is fertilised.

      .... [1]

   c. Match each of the following terms with the appropriate definition.

   AID   conception   copulation   development   fertilisation   implantation

Definition	Term
The fusion of male and female gametes	
The beginning of development of a new individual	
Sexual intercourse	
Fertilisation using sperm from a donor male	
The attachment of the fertilised egg to the lining of the uterus	
The stages which occur as cells divide and become organised into tissues and organs	

   [5]

Extension

2. A woman gave birth to triplets, two identical sons and a daughter. Describe the events that occurred at the time of fertilisation or shortly afterwards to result in this multiple birth.

3. Remind yourself of the definition of fertilisation. Suggest how a female gamete prevents 'fertilisation' by more than one male gamete.

69

Human reproduction — 3.9 Contraception

1. a. State two reasons for using a condom during sexual intercourse.

   ....

   .... [2]

   b. The table below shows some methods of contraception and their rates of failure.

Method	Rate of failure	Ranking
Condom	1 in 7	
Coil or IUD	1 in 20	
Diaphragm	1 in 8	
Pill	1 in 300	
Rhythm	1 in 4	
Sterilisation	1 in 30 000	

   i. Complete the rank order in the table. The best method, i.e. lowest rate of failure, scores 1 and the one with the greatest rate of failure scores 6. [1]

   ii. Explain how the contraceptive pill prevents pregnancy.

   ....

   ....

   ....

   .... [3]

   iii. If 1000 women were using the IUD calculate how many would become pregnant after one year. Show your working.

   .... [2]

2. **Extension** It is now possible in some countries to receive 'morning after' contraception. Suggest one advantage and one disadvantage of this.

Human reproduction   3.10 Placenta

1. The diagram below shows a human uterus containing a developing fetus.

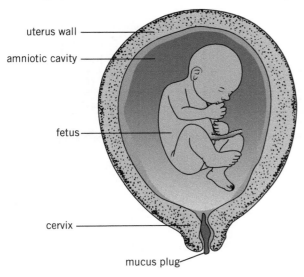

a. i. Name two structures shown in the diagram which protect the fetus.

   For each structure you have named, explain how it carries out its protective function.

   Structure 1: name .... [1]

   Explanation of function ....

   .... [1]

   Structure 2: name .... [1]

   Explanation of function ....

   .... [1]

   ii. Draw in the position of the placenta on this diagram. [1]

b. Complete this table by placing a tick (✓) in ONE box on each line to show the movement, if any, of each substance across the placenta.

Substance	From mother to fetus	From fetus to mother	No movement across placenta
Glucose			
Haemoglobin			
Nicotine			
Amino acids			
Carbon dioxide			
Alcohol			
Urea			

[7]

Extension

2. Humans are placental mammals. What are monotremes and marsupials? Explain how they differ from placental mammals.

Human reproduction
3.11 Feeding babies

1. The table below compares the minimum daily requirements of infants (under 18 months old) and adults (over 18 years old).

	Energy requirements (kJ)	Protein (g)	Iron (mg)	Calcium (mg)	Vitamin D (µg)
Infant	2400	15	6	650	10
Adult	8800	45	10	450	2.5

   a. Calculate the percentage of the adult's energy requirement required by the infant. Show your working.

   ....% [2]

   b. Explain why the infant requires more calcium and vitamin D than the adult, but less iron.

   More calcium and vitamin D because ....

   .... [1]

   Less iron because ....

   .... [1]

   This table summarises the composition of 1 kg of cow's milk.

	Energy content (kJ)	Protein (g)	Iron (mg)	Calcium (mg)	Vitamin D (µg)
Cow's milk	2750	35	1	1200	0.3 (summer) 0.15 (winter)

   c. Calculate how much cow's milk an infant would need to satisfy its energy requirement. Show your working.

   .... [2]

   d. Use data from the two tables to suggest a disadvantage of feeding an infant on cow's milk.

   ....

   .... [2]

   e. Suggest a reason why cow's milk contains more vitamin D in summer than in winter.

   .... [1]

Extension

2. Once a woman becomes pregnant she will be offered advice on how to care for her developing baby. Suggest why expectant mothers are advised to (a) reduce intake of alcohol, (b) stop smoking, and (c) increase intake of dairy products.

Human reproduction — 3.12 Birth

1. At the end of the period of development, a sequence of events leads to the birth of the baby. The final stage is called labour, and begins with the contraction of the uterus muscle.

   Complete this paragraph to describe which hormones are involved in controlling this process.

   Choose words from the following list. Words may be used once, more than once, or not at all.

   adrenaline   falls   oestrogen   oxytocin   progesterone   rises   testosterone

   The contractions of the uterus are prevented by ...., and the concentration of this hormone .... as birth approaches.

   The contractions are stimulated by ...., a hormone which also stimulates lactation.

   The contractions are helped by ...., so the level of this hormone .... as birth approaches.

2. a. A baby is about to be born after a full-term pregnancy.

   i. State the name of the period between implantation of the zygote and birth.

   .... [1]

   ii. Suggest how long this period is in most women. Choose your answer from the following alternatives.

   one month   42 weeks   one year   38 weeks   [1]

   b. i. Suggest why it is better for the baby to be born head first.

   ....
   .... [1]

   ii. State how the baby is pushed out of the uterus at birth.

   ....
   .... [1]

Extension

3. Suggest why breastfeeding can act as a natural method of contraception.

Human reproduction

3.13 Sexually transmitted infections

1. Gonorrhoea is a sexually transmitted infection (an STI).

   a. i. State which **type** of organism causes gonorrhoea.

      .... [1]

      ii. State **one** early symptom of gonorrhoea.

      .... [1]

      iii. State which method of contraception is most likely to prevent transmission of this disease.

      .... [1]

   b. Another STI is AIDS.

      i. Explain how AIDS affects the human body.

      .... [2]

      ii. Explain why AIDS and gonorrhoea cannot be treated by the same drugs.

      .... [2]

Extension

2. Scientists have discovered that some women in East Africa do not become infected with AIDS. Suggest why these women are protected, and how this information might lead to better treatment for the disease.

Inheritance — 3.14 Variation and inheritance

1. Tongue rolling is an example of discontinuous variation. It is partly controlled by a dominant allele, R, of a single gene.

   a. i. Name one other example of discontinuous variation in humans.

   .... [1]

   ii. Define the term *allele*.

   .... [1]

   iii. Draw a simple labelled diagram to show how genes and chromosomes are related to one another. [2]

   b. This diagram shows a family history (a pedigree) of tongue rolling over three generations.

   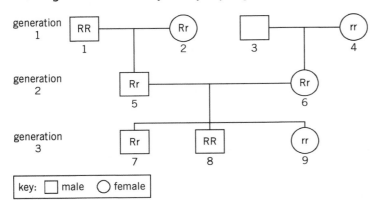

   i. State which individual in generation 3 is *not* a tongue roller. .... [1]

   ii. Suggest the possible genotypes of individual 3. .... [1]

   iii. Individual 7 marries a woman who is heterozygous for tongue rolling.

   State the probability (chance) that their second child will be a tongue roller.

   .... [1]

   c. Scientists often trace the inheritance of characteristics through mitochondrial DNA. Explain what is meant by mitochondrial DNA, and suggest why it provides evidence of inheritance through the maternal line only.

Inheritance
3.15 DNA and characteristics

1. The diagram shows a short sequence of bases in a DNA molecule. The base sequence in DNA carries information in the form of a genetic code.

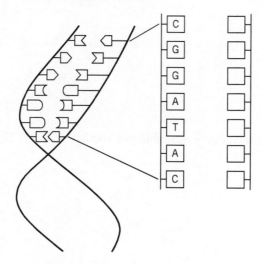

   a. Write down the sequence of bases on the other strand of the DNA molecule. [2]

   b. A modified copy of the DNA sequence is carried to the cytoplasm of the cell.

   i. State the name of this modified sequence. .... [1]

   ii. State the name of the structure in the cell where the code is 'read' by the cell. .... [1]

   c. A sequence of three bases carries a code for another type of biological molecule.

   State the name of this type of molecule. .... [1]

   d. The type of molecule named in your answer to part c. is built up into larger molecules, and these molecules determine the characteristics of organisms.

   i. State the general name given to these larger molecules. .... [1]

   ii. Complete this table to match the molecules to their functions (i.e. to the characteristics they give to an organism). [5]

Molecule	Characteristic
	Ability of red blood cells to transport oxygen
	Bind to and identify molecules on the surface of invading microbes
Receptor protein in synapse	
Lipase	
	Provides strength and structure to hair and nails

76

Inheritance   **3.16 DNA and how the genetic code is carried**

1. Scientists have found that they can use small rings of DNA called plasmids to transfer useful genes into bacterial cells. The bacterial cells can then manufacture the protein coded by this gene. The protein can be extracted from the bacterial cells, and may be very valuable in medical treatments for humans.

   a. One useful protein made by genetic engineering can be used to break down starch during food production.

      State the name of this protein. .... [1]

   b. Some people believe that genetic engineering can be dangerous, but not everyone agrees with this.

      Suggest two advantages of, and two possible concerns about, genetic engineering.

      Two advantages ....

      .... [2]

      Two possible concerns ....

      .... [2]

Extension

2. One important protein made by mammals (including humans) is insulin. How many amino acids are present in a molecule of insulin? Calculate how many base pairs will be necessary to code for one insulin protein molecule.

3. Find out how many base pairs are found in one set of human chromosomes. Count how many letters are found on one page of your biology textbook, and then calculate how many pages you would need to write down all of the base letters in one set of human chromosomes.

Inheritance — 3.17 Cell division

1. Human body cells usually contain 23 pairs of chromosomes. The exceptions to this rule are the gametes and the mature red blood cells.

   a. Complete the table below.

Type of cell	Total number of chromosomes	Type of sex chromosomes present
Male nerve cell		
Female white blood cell		
Sperm cell		
Egg cell/ovum		
Red blood cell		

   [5]

   b. Blood cells are produced in bone marrow. State the name of the type of cell division which produces them.

   .... [1]

2. This diagram shows some stages in cell division.

   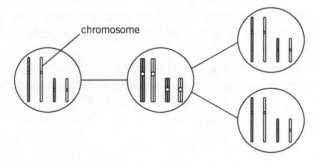

   a. State the diploid number of chromosomes in these cells. .... [1]

   b. Name the type of cell division shown. .... [1]

   c. Give **two** examples of sites where this type of cell division occurs.

   1. ....

   2. .... [2]

3. Methotrexate is a drug used to slow down the division of some human cells, for example in the treatment of some cancers. Suggest why doctors insist on regular blood tests for people being treated with this drug.

Inheritance — 3.18 Inheritance

1. This paragraph describes some features of inheritance.

   Use words from this list to complete each of the spaces in the paragraph. Each word may be used once, or not at all.

   | allele | diploid | discontinuous | dominant | fertilisation | gene | haploid |
   | heterozygous | homozygous | meiosis | mitosis | recessive | redundant | |

   In humans, eye colour may be brown or blue. This is controlled by a single .... which has two forms.

   Gametes are formed by the type of cell division called .... : the gametes are

   .... and fuse at .... to form a

   .... zygote.

   Two humans both have brown eyes, but one of their three children has blue eyes. This means that blue eye colour is

   controlled by a .... allele and that both of the parents are .... [7]

2. Study the two lists below. One is a list of genetic terms and the other is a list of definitions of these terms.

   Draw guidelines to link each term with its correct definition.

Genetic term
Genotype
Homozygous
Dominant
Heterozygous
Recessive
Chromosome
Allele
Phenotype

Definition
The observable features of an organism
An allele that is always expressed if it is present
Having two alternative alleles of a gene
One alternative form of a gene
The set of alleles present in an organism
Having two identical alleles
A thread-like structure of DNA, carrying genetic information in the form of genes
Two identical alleles of a particular gene

   [8]

Extension

3. Some closely related mammals, such as a horse and a donkey, can produce offspring. Suggest why these offspring are very rarely fertile.

Inheritance
3.19 Patterns of inheritance

1. A set of triplets was born, but their mother died during the birth. The babies were separated from one another, and brought up by different families. When they were 18 they met up with one another for the first time since their separation. Various measurements were made on them, and some of the information obtained is recorded in this table.

	Andrew	John	David
Mass (kg)	81	92	86
Height (cm)	179	180	179
Blood group	O	O	AB
Intelligence quotient (IQ)	128	138	140

   a. i. State which two of the boys might be identical twins from this evidence.

   .... and .... [1]

   ii. Explain which piece of evidence is most important in helping you to reach this conclusion.

   ....

   .... [1]

   iii. Suggest a reason why Andrew and David have such different body mass measurements although they are of the same height.

   .... [1]

   b. The ABO blood group is determined by three alleles, although only two are present in any one cell. The three alleles are given the symbols I^A, I^B, and I^O.

   The relationship between genotype and phenotype for these blood groups is shown in the table below.

Genotype	$I^A I^A$	$I^A I^O$	$I^B I^B$	$I^B I^O$	$I^A I^B$	$I^O I^O$
Phenotype	A	A	B	B	AB	O

   State the name of this type of genetic relationship, in which a particular combination of alleles (in this case $I^A I^B$) results in a new phenotype.

   .... [1]

> **Extension**
>
> 2. Draw a Punnett square to suggest how two pink-flowered plants can produce red, pink, and white flowers amongst their offspring.

Inheritance

3.20 Inherited medical conditions

1. Cystic fibrosis is an inherited disorder that affects the production of mucus and sweat. The mucus is not liquid enough, so is very sticky. It can block air passages in the lungs, and also the ducts from the pancreas and, in males, the testes.

   The condition is caused by a recessive allele.

   The diagram shows a family tree.

   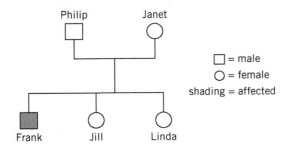

   a. Suggest appropriate symbols for the dominant allele and for the recessive allele.

   Dominant .... Recessive .... [1]

   b. Use these symbols to give the genotype of the two parents.

   Philip .... Janet .... [2]

   c. State the possible genotypes of i. Frank .... and

   ii. Linda .... [2]

   d. If Philip and Janet decide to have another child, what is the probability that it will be affected by cystic fibrosis?

   .... [1]

   e. The pancreas releases a lipase and several proteases.

   Explain why an affected child often grows and develops more slowly than a child without the condition.

   ....
   ....
   .... [3]

2. Find out about the patterns of inheritance for **Tay-Sachs disease** and for **achondroplasia**.

Extension

Inheritance

3.21 X and Y chromosomes

1. This diagram shows the arrangement of chromosomes in a cell of a human fetus.

   [Karyotype diagram showing chromosome pairs 1–22 and X]

   a. State the gender (sex) of this fetus. .... [1]

      Explain your answer. .... [1]

   b. The fetus has a chromosome mutation, which may lead to a genetic abnormality. Use the diagram to identify this chromosome mutation.

      .... [1]

2. Haemophilia is a sex-linked inherited disease. It is controlled by alleles carried on the X chromosome (**H** for normal clotting, **h** for haemophilia).

   The table below gives some information about possible combinations of phenotype and genotype.

   Complete the table by:

   a. drawing the correct chromosomes in boxes 2, 3, and 5

   b. adding the symbols for the alleles to the correct positions on the chromosomes

   c. writing into the final row the sex of each individual.

   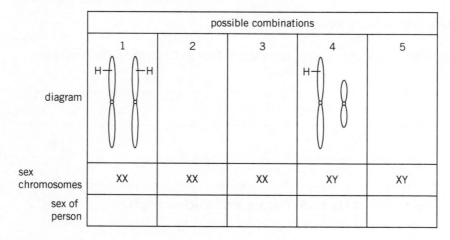

   [5]

3. Very rarely the separation of X and Y chromosomes at meiosis does not take place properly, and a gamete may receive an unusual number of sex chromosomes. Suggest the likely characteristics of an individual with an XYY genotype. Suggest why lawyers may use an XYY genotype as a reason for some criminal behaviour.

Variation and selection   3.22 Variation

1. Study the two lists below. One is a list of terms relating to variation and the other is a list of definitions of these terms.

   Draw guidelines to link each term with its correct definition.

Genetic term		Definition
Continuous variation		The observable features of an organism
Gene		A form of variation with many intermediate forms between the extremes
Discontinuous variation		One possible form of environmental influence on variation
Phenotype		One example of discontinuous variation
Height in humans		This factor, in addition to genotype, can affect phenotype
Environment		A section of DNA responsible for an inherited characteristic
Nutrients		One example of continuous variation
Blood group		A form of variation with clear-cut differences between groups

[8]

Extension

2. Actors sometimes need to change their appearance in order to play a part in a film production. Both hair and eye colour are frequently altered to fit into a specific role. Use the terms phenotype and genotype to explain how they might make these changes.

83

Variation and selection   3.23 Causes of variation

1. a. Sickle cell anaemia is inherited. It is caused by a recessive allele (**Hb$_S$**): the normal red blood cell allele is **Hb$_A$**.

Study the family tree (below).

key:
- ○ female not affected
- ● female carrier
- ⊘ female, condition unknown
- □ male not affected
- ■ male carrier
- ? male, condition unknown

Family tree: Lucy — Peter; their children Ann, Alice, Robert; Michael married Ann (child Tom ?); Alice; Robert married Jane (children James ?, Jo ?).

   i. State Lucy's genotype. .... [1]

   ii. Suggest which person in the family tree could have sickle cell anaemia. Explain your answer.

   Name ....

   Explanation ....

   .... [2]

b. A person heterozygous for sickle cell anaemia (**Hb$_A$ Hb$_S$**) shows some symptoms of anaemia. These may lead to weakness and tiredness, although they are rarely severe enough to be fatal. Despite these disadvantages there are situations in which the sickle cell allele may give an advantage to the carrier.

Explain how the **Hb$_S$** allele may be a selective advantage in certain environments.

....

....

.... [2]

2. Charles Darwin believed that new **species** can arise by natural selection of variations within a population. Since Darwin's time, scientists have suggested that variation may arise by **mutation**, as well as by the formation of new combinations of **genes** at fertilisation during sexual reproduction.

   a. State the meanings of the terms:

   i. species .... [1]

   ii. mutation .... [1]

   iii. gene .... [1]

   b. Suggest **two** factors that may increase the rate of mutation in a population.

   i. ....

   ii. .... [2]

Extension

3. Gametes can differ from one another as a result of crossing-over and independent assortment. Ignoring crossing-over, calculate how many different male gametes could be formed in a mature human male (big clue – 2 pairs of chromosomes provide 4 possible different gametes).

Variation and selection   3.24 Selection

1. a. The table below shows some examples of selection.

Example	Type of selection: natural or artificial
Cattle being bred that are able to withstand cold winters	
The development of a strain of rice that is more resistant to disease	
The resistance of a strain of bacteria to a particular antibiotic	
The ability of a species of shrub to grow on soil containing high amounts of copper	
The similarity between a moth's wing pattern and its habitat, making it less conspicuous to predatory birds	

   Complete the table to show which type of selection, natural or artificial, is involved. [5]

   b. The bar graph shows the average milk yield for a herd of Friesian cows between 1976 and 2004.

   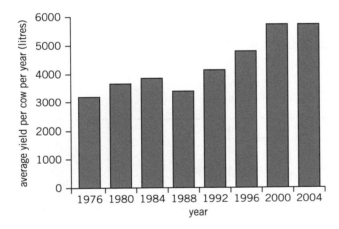

   i. Explain how animal breeders might have caused the trend between 1976 and 2004 to occur.

   ....

   ....

   .... [4]

   ii. Apart from the yield of food (milk or meat), suggest another characteristic that farmers might improve in animals to make a farm more economically successful.

   .... [1]

2. One animal studied by Charles Darwin was the Great Spotted Woodpecker (*Dendrocopos major*). Make a drawing to show the adaptations of this bird to its way of life.

85

Variation and selection 3.25 Natural selection

1. The Everglades is an enormous flooded area in southern Florida, USA.

   There are many different species of animals and plants there, and they are well adapted to their environment.

   a. The animals and plants have become well adapted by the process of natural selection.

      The following are stages in the evolution of species by means of natural selection.

      A survival of the fittest

      B over-production of offspring

      C competition causes a struggle for existence

      D advantageous characteristics are passed on to offspring

      E variation occurs between members of the same population

      Use the letters **A–E** to rearrange these stages into the correct sequence.

      …..   …..   …..   …..   ….. [5]

   b. Some of the early European explorers in Florida noticed that there were many species of bird.

      The heads and beaks of birds are often well adapted to their diet. The diagrams in this table show the heads of some species of birds.

      Draw guidelines to match up the heads with the likely diet of the birds.

| Feeds on nuts and other hard fruits | Filters algae and other small organisms from the water | Feeds by spearing fish and frogs | Captures Florida rabbits and other mammals | Feeds by catching small insects |

[5]

Extension

2. Find out who Trofim Lysenko was. How did his ideas on evolution differ from those of Charles Darwin?

Variation and selection — 3.26 Artificial selection

1. The diagrams below show the effect of selective breeding on wild cabbage (*Brassica oleracea*).

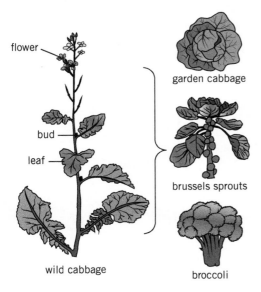

a. State which part of the wild cabbage has been selected for when breeding the:

   brussels sprouts ....

   broccoli .... [2]

b. Once a plant has been selected as a new variety of broccoli, the grower then produces many more identical plants by taking cuttings.

   i. Explain why the cuttings grow to be identical to the parent plant.

   ....

   ....

   .... [2]

   ii. The cuttings are usually planted into a type of compost containing plant hormones (to encourage rooting) and additional nitrate. The cuttings are kept in a closed transparent enclosure.

   Explain the importance of:

   nitrate in the growth compost ....

   .... [1]

   keeping the plants in a closed environment. ....

   .... [1]

2. **Extension** — Make an internet search to find an image (or group of images) that could be used to illustrate how artificial selection could produce a Springer Spaniel from an ancestral wolf.

Ecosystems, decay, and cycles

4.1 Ecology and ecosystems

1. A student used a 50 cm quadrat to investigate the distribution of small insects called springtails in the leaf litter of a woodland. The number of springtails in each quadrat is shown in the diagram.

0	15	0	0	1
1	1	0	0	3
0	0	2	1	2
3	3	12	14	0
1	0	0	0	1

   50 cm × 50 cm

   a. Calculate the area of each quadrat.

   .... m^2 [1]

   b. Calculate the mean number of springtails per square metre in the area studied by this student. Show your working.

   Mean number of springtails per square metre = ... [2]

   c. The springtails were not evenly distributed in the woodland. They were found mainly under pieces of dead wood.

   Suggest **two** advantages to springtails of living underneath the pieces of dead wood.

   i. ....

   ii. .... [2]

> **Extension**
>
> 2. Zookeepers try to copy an animal's environment as closely as possible when they design a modern zoo enclosure. Suggest what should be included in their design if they are to try to breed animals for release into the wild.

Ecosystems, decay, and cycles

4.2 Ecology and the environment

1. Feeding relationships

   Use guide lines to match the terms with their definitions.

Term/process		Definition
Food chain		An organism that gets its energy from dead or waste organic material
Food web		An organism that gets its energy by feeding on other organisms
Producer		An animal that gets its energy by eating other animals
Consumer		A network of interconnected food chains
Herbivore		The transfer of energy from one organism to the next, beginning with a producer
Carnivore		An organism that makes its own organic nutrients, usually through photosynthesis
Decomposer		An animal that gets its energy from eating plants

2. A group of students studied a food web for an English lake.

   Algae are small microscopic plants. Water fleas are small crustaceans about 2 mm in length. Smelt are fish – they are adult at about 4 cm in length. The kingfisher is a bird, about 22 cm long.

   Draw and label a likely pyramid of numbers for the food chain linking:

   algae ⟶ water fleas ⟶ smelt ⟶ kingfisher

   [2]

Extension

3. An efficient plant can convert about 10% of the light energy that falls on it into chemical energy. Find out the efficiency of a solar panel in the conversion of light to electrical energy.

Ecosystems, decay, and cycles

4.3 Energy flow

1. a. This diagram represents the energy flow through a food chain.

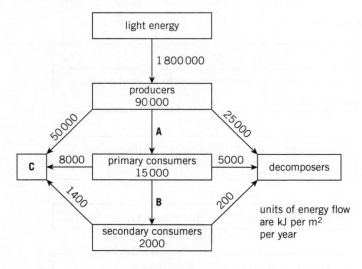

i. Name the source of light energy. .... [1]

ii. State the process occurring at **A** and **B**. .... [1]

iii. Explain why not all of the light energy available is absorbed by the producers.

....

.... [2]

iv. State the name of the process shown at **C**, and the form in which the energy is lost.

Process .... [1]

Form of energy lost .... [1]

b. i. Calculate the percentage of energy transferred at stages **A** and **B**.

Show your working.

A .... B .... [2]

ii. Suggest how evidence in the diagram supports the idea that humans should eat more vegetable matter and less meat.

....

....

.... [3]

> **Extension**
>
> **2.** What is a **pyramid of energy**? Suggest why an ecologist would be more interested in a pyramid of energy than in a pyramid of biomass.

Ecosystems, decay, and cycles

4.4 Decay

1. A group of students decided to investigate the decay of leaves.

   Four samples of equal masses of leaves were placed in bags made of plastic mesh.

   The mesh size in bags **A**, **B**, and **C** was the same, but the mesh size in bag **D** was much smaller.

   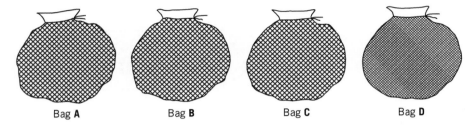

   Bag **A**   Bag **B**   Bag **C**   Bag **D**

   - The bags of leaves were weighed.
   - The bags were buried at the same depth in different soils.
   - The bags were kept at the same temperature.
   - After 100 days the bags were dug up and reweighed.

   The results of the investigation are shown in this table.

Bag	Location	Type of soil	Mass at start (g)	Mass after 100 days (g)	Loss of mass (g)
A	Garden	Wet clay	60	58	2
B	Woodland	Dry, sandy	60	51	9
C	Woodland	Moist	60	32	
D	Woodland	Moist	60	43	

   a. i. Complete the table to show the loss in mass in bags **C** and **D**. [1]

   ii. Explain how the results suggest that decay requires water and air (oxygen) to take place.

   ....

   .... [2]

   iii. Explain the difference between the results for bags **C** and **D**.

   .... [1]

   iv. Explain why it was important that the bags were made of plastic.

   .... [1]

   b. State the names of **two** types of organism responsible for decay.

   .... [2]

Extension

2. Wood is a natural product, and has been used for more than a hundred years in the manufacture of railway sleepers and telegraph poles. Railway builders in East Africa found that wooden products were quickly destroyed by both termites and decomposing microbes. Suggest how humans can protect wood against decomposers.

Ecosystems, decay, and cycles

4.5 The carbon cycle

1. The first part of this question asks you to use a list of words to fill in a set of boxes. This is quite a common way to test your knowledge and understanding of a biological process. The diagram below represents the carbon cycle.

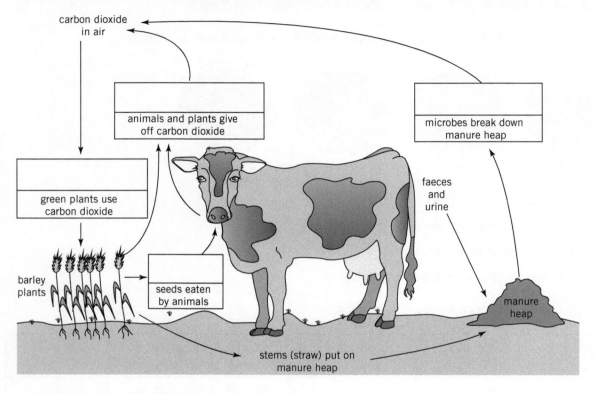

   a. Use words from this list to complete the boxes in the diagram.

   combustion   decay   excretion   feeding   photosynthesis   respiration   [4]

   b. In some countries the manure heap is collected, dried, and then burned as a fuel.

   State the effect that this would have on the carbon dioxide concentration in the air.

   .... [1]

Extension

2. Commercial businesses now often advertise that they plant trees to reduce their carbon footprint. Explain what is meant by a carbon footprint. Calculate the carbon footprint of your families' transport requirements (cars and aeroplanes) over the past year.

Ecosystems, decay, and cycles

4.6 The nitrogen cycle

1. a. The diagram shows the nitrogen cycle.

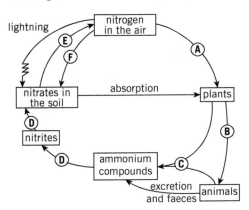

   i. Name the processes **B** and **E**.

   B ....

   E .... [2]

   ii. Name the process and the organisms involved in **C**.

   Process ....

   Organisms .... [2]

   b. A heap of dead leaves had been left to decompose. Part of the decomposition sequence is shown below.

   dead leaves → (decomposers) → ammonium → (*Nitrosomonas* bacteria) → nitrite → (*Nitrobacter* bacteria) → nitrate

   Use this sequence and your own knowledge to answer the following questions.

   i. Suggest **three** effects on the sequence if the *Nitrosomonas* bacteria died out.

   1. ....

   2. ....

   3. .... [3]

   ii. Tick **two** boxes in this table to show factors which would speed up the decomposition process.

Factor	Box
Maintaining an acid pH	
A plastic cover to exclude air	
Few scavenging insects	
Many scavenging insects	
Regular turning of the heap to add air	

   [2]

2. Explain how root nodules illustrate the principle of symbiosis.

Extension

93

Ecosystems, decay, and cycles

4.7 The water cycle

1. a. The diagram below summarises the water cycle.

   Fill in the boxes using words from the list. You may use each word once, more than once, or not at all.

   drainage   egestion   evaporation   condensation
        photosynthesis   rainfall   transpiration   melting/freezing

   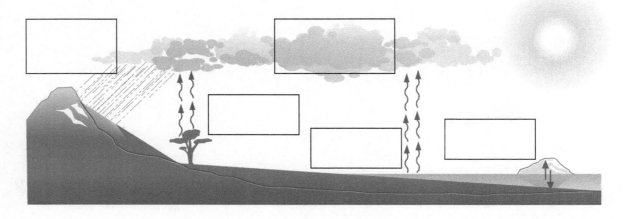

   [5]

   b. i. State **two** ways in which water may leave the body of a mammal.

   1. ....

   2. .... [2]

   ii. Describe how water in a plant may become water in the cells of an animal.

   ....

   ....

   ....

   ....

   ....

   .... [3]

Extension

2. Explain why space scientists are so excited about the possibility of ice on the 'dark' side of the Moon.

94

Populations — 4.8 The size of populations

1. This diagram shows a number of factors that can affect the size of a population.

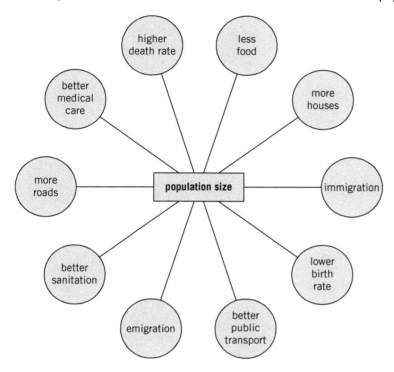

   a. State **four** factors, shown in the diagram, that reduce the size of a population.

   ....

   ....

   ....

   .... [4]

   b. Suggest **two** ways in which better medical care can lead to an increase in the size of a population.

   ....

   .... [2]

   c. State **two** ways in which humans control bacterial populations. [1]

   ....

   .... [2]

Extension

2. The rose-ringed parakeet (*Psittacula krameri*) is an 'alien' to Britain. The population in the wild was probably no more than 30–50 in the 1980s. What is its population now? Explain why its population has increased so rapidly.

95

Populations — 4.9 Human population changes

1. a. Explain what is meant by the term **age structure** of a population.

....

....

.... [2]

b. List and explain **three** factors that can affect the age structure of a population.

1. Factor .... Explanation of effect ....

....

....

2. Factor .... Explanation of effect ....

....

....

3. Factor .... Explanation of effect ....

....

.... [6]

Extension

2. Explain why the population of Japan is not growing, whereas the population of Brazil is increasing rapidly. Explain why the population of Western Europe is growing even though the birth rate is not increasing significantly.

Microorganisms

4.10 Bacteria in biotechnology

1. a. For each of the following statements, mark as True (T) or False (F).

	Statement	True or False
1	A typical bacterium is about one thousandth of a metre wide	
2	Bacteria are smaller than viruses	
3	Gonorrhoea is caused by a bacterium	
4	Bacteria may contain plasmids	
5	All bacteria are harmful	
6	Bacteria multiply by binary fission	
7	A bacterium called *Vibrio* causes cholera	
8	Bacteria can produce the enzyme protease	
9	Bacteria can be genetically modified to produce human insulin	
10	Some bacteria can produce their own organic chemicals using energy from the Sun	
11	Bacteria in the human intestine produce some of the vitamins required for human health	
12	Bacteria have a nucleus, but it is smaller than a human nucleus	

[12]

b. For any three of the statements which you have labelled as False, explain why you have made that choice. [1]

1. Statement ....

   Explanation ....

   .... [2]

2. Statement ....

   Explanation ....

   .... [2]

3. Statement ....

   Explanation ....

   .... [2]

Extension

2. It is theoretically possible that a single bacterium in the human gut could become 2^{72} cells in a 24-hour period. Suggest why this doesn't happen.

Microorganisms

4.11 Biotechnology: the production of penicillin

1. The diagram shows a bioreactor used for the production of penicillin.

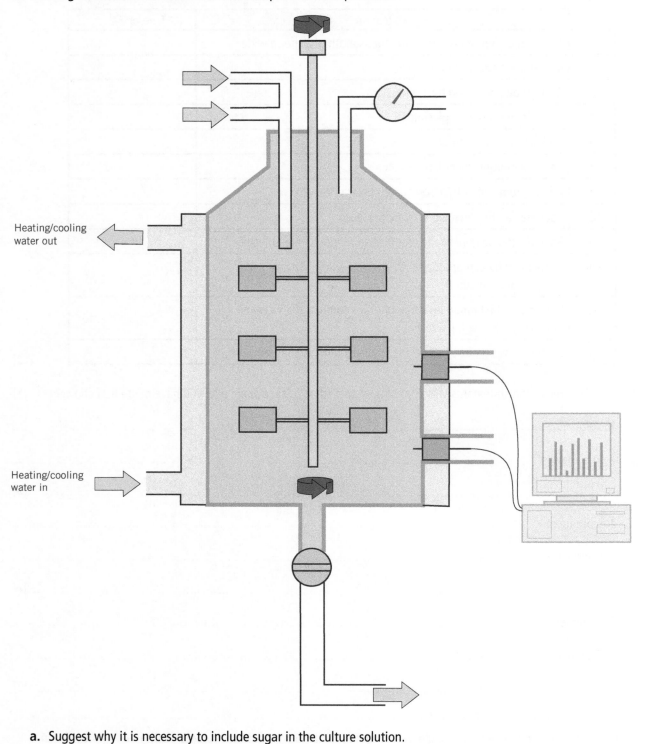

a. Suggest why it is necessary to include sugar in the culture solution.

.... [1]

b. It is necessary to maintain a constant temperature inside the bioreactor. Explain **why** this is necessary.

....

.... [2]

Microorganisms
4.12 Biotechnology: the production of penicillin

c. This graph shows the amounts of *Penicillium* fungus and penicillin in the bioreactor over a period of nine days.

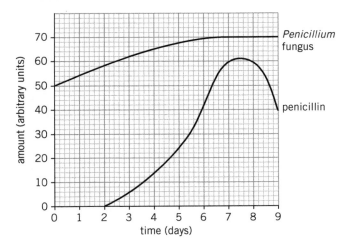

Suggest the best time to collect the penicillin.

....

Explain your answer.

....

.... [2]

d. Doctors are reluctant to prescribe penicillin for all illnesses.

Explain how the overuse of antibiotics can lead to the development of resistant strains of bacteria.

....

....

....

....

.... [3]

e. Bioreactors may also be used in the production of enzymes.

Complete this table about commercially valuable enzymes. One section has been completed for you.

Name of enzyme	Commercial value
	Clearing fruit juices by digesting clumps of plant cells
Lipase from fungi	Improves chocolate flow when coating biscuits
	Part of biological washing powders – remove blood stains
Lactase	

f. What is MRSA? Explain why people are so concerned about MRSA. Suggest what people can do to limit the effects of MRSA.

Extension

Microorganisms 4.13 Yeast

1. Scientists in the brewing and baking industries are interested in the metabolism of yeast.

   A group of scientists investigated the production of alcohol from four different sugars, **M, N, O,** and **P**. They set up apparatus as shown in diagram **A** for each of the different sugars. The length of the carbon dioxide bubble shown in diagram **B** was measured at five-minute intervals.

   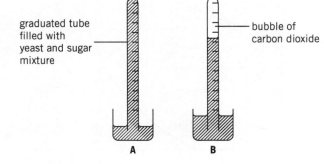

   - In each case the same concentration of sugar was used.
   - The temperature was kept at 25 °C throughout the experiments.
   - This process is called alcoholic fermentation.

   a. i. Explain the purpose of the yeast and sugar mixture. ....

   .... [1]

   ii. State the independent and dependent variables in this set of experiments.

   independent ....

   dependent .... [2]

   iii. Explain why it is important that the tubes were **completely** filled with the mixture at the start of the experiments. ....

   .... [1]

   iv. Suggest why the rate of carbon dioxide production slowed down after about 25 minutes. ....

   .... [2]

   b. Write a word equation for the process of alcoholic fermentation.

   [2]

   c. Suggest **two** ways in which the rate of alcohol production could have been increased.

   ....

   .... [2]

Extension

2. What are *Saccharomyces carlsbergensis* and *Candida utilis*? Explain why scientists would like to understand how to control the growth of these two organisms.

Human impact on ecosystems

4.14 Genetic engineering

1. a. The following table gives a list of some terms associated with genetic engineering, and some definitions for these terms.

   Match the terms in column 1 with the correct definition from column 2.

Term
Gene
Plasmid
Vector
Ligase
Restriction
Sticky ends
Fermenter

Definition
A small circle of DNA in a bacterial cell
An enzyme that can splice one gene into another section of DNA
An enzyme that can cut out a specific gene from a chromosome
Pieces of single-stranded DNA left exposed after a gene is removed from a chromosome
A section of DNA coding for a protein
A vessel in which engineered bacteria can produce a valuable product under optimum conditions
A structure which can carry a gene into another cell

   [6]

   b. Genetic engineering can be used to produce protein products that are useful to humans.

   State **one** reason why each of the following proteins is useful to humans.

   1. Insulin ....

   ....

   2. Factor 8 ....

   ....

   3. Pectinase ....

   ....

   4. Human growth hormone ....

   .... [4]

   c. Before the development of genetic engineering, vaccines were made by heating or chemically treating whole viruses. The vaccine therefore contained whole virus particles.

   Explain why it is safer to use a genetically engineered vaccine rather than one made directly from the hepatitis B virus.

   ....

   ....

   .... [2]

> **Extension**
>
> 2. Explain how gene therapy may provide a treatment for retinal macular degeneration.

101

Human impact on ecosystems

4.15 Humans and agriculture

1. The table below shows the energy content and the concentration of insecticide in a group of organisms in a food chain.

Organism in food chain	Energy content (percentage of original energy)	Concentration of insecticide in body (mg per kg)
Human	1	1.5
Food fish	3	0.1
Small fish	9	0.03
Microscopic animals	20	0.01
Algae (microscopic plants)	100	0.001

a. Calculate the percentage of energy loss between the microscopic plants and the food fish.

.... % [1]

b. Using information from the table, explain why food chains involving humans should be kept as short as possible.

....

.... [2]

c. Describe one way in which insecticides have been harmful to organisms other than the target insects.

....

.... [2]

> **Extension**
>
> 2. What is a molluscicide? Explain how the spread of one disease-causing organism can be controlled by the use of a molluscicide.

Human impact on ecosystems

4.16 Land use for agriculture

1. More food and raw materials are required to meet the demands of an increasing world population.

   Describe and explain the possible harmful effects on organisms and the environment of burning large areas of Indonesian rainforest, so that the land can be used to grow palm oil trees.

   ....

   ....

   ....

   .... [4]

2. These diagrams show changes to a farm between 1953 and 2003.

   key
   hedges ——— boundary -----
   river ～～～ buildings ■■
   trees ✤ ✤ marsh ⁎ ⁎ ⁎

   The fields on the farm are separated by hedges.

   a. i. State **two** major changes which were made to the land between 1953 and 2003.

      1. ....

      2. .... [2]

   ii. Suggest and explain **two** ways in which these changes would affect wildlife on the farm.

      1. ....

         ....

      2. ....

         .... [4]

   b. Farmers often remove areas of woodland to provide more space for growing crops.

      Suggest three disadvantages of this deforestation.

      1. ....

      2. ....

      3. .... [3]

Extension

3. A recent survey of the grey partridge (*Perdix perdix*) in Britain found that its population was declining rapidly. Ecologists suggested that this was due to agricultural practices used in the UK. Explain how modern agriculture might have caused this population decline. Suggest how it could be reversed.

Human impact on ecosystems

4.17 Malnutrition and famine

1. a. This set of pie charts shows the food groups present in eggs, milk, rice, and beans.

   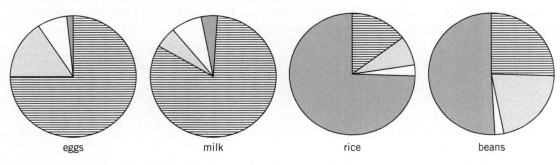

   eggs　　　　milk　　　　rice　　　　beans

   key to food groups
   ▤ water　☐ proteins　☐ fats and oils　▦ carbohydrates

   A malnourished person may have an unbalanced diet that may be missing some essential nutrients.

   i. State which food has the highest concentration of protein. .... [1]

   ii. State which food has the most carbohydrate. .... [1]

   iii. Name **two** other food groups not shown in the pie charts.

   ....

   .... [2]

   b. An extreme and continued shortage of food can result in a famine.

   State **three** factors which may cause this shortage of food.

   1. ....

   2. ....

   3. .... [3]

Extension

2. Some people suggest that wealthy countries should not provide food for people living in desert areas, as these areas can never supply enough food to support a large human population. These people say that this food support will have to go on indefinitely, and that the population should be allowed to 'balance' to what the environment can provide. What do you think?

Human impact on ecosystems

4.18 Pollution

1. The table below provides some data about a number of atmospheric gases.

Name of gas	Source(s) of gas	Percentage influence on greenhouse effect
CFCs	Air conditioning systems; aerosol propellant; refrigerators	13
Carbon dioxide	Burning forest trees; burning fossil fuels; production of cement	56
Nitrous oxide	Breakdown of fertilisers	6
Methane	Waste gases from animals such as sheep, cattle, and termites; rotting vegetation	25

   a. Explain why these gases are called 'greenhouse' gases.

   ....

   ....

   .... [2]

   b. From this data suggest why the following practices should be encouraged:

   1. Development of renewable energy resources such as wind turbines.

      .... [1]

   2. Improved insulation of walls and roofs in houses.

      .... [1]

   3. Reforestation.

      .... [1]

2. There are large copper mines in Tanzania. One of these mines is so far from railway links that all supplies must be brought in, and all products and waste materials removed, by trucks. These trucks run throughout the day and night.

   Describe and explain the possible harmful effects on organisms and the environment.

   ....

   ....

   .... [3]

Extension

3. Recent research has shown that some car manufacturers have been providing false information about the emission of PM. Explain what PM is, and why it is dangerous to health.

   ....

105

Human impact on ecosystems
4.19 Eutrophication

1. Organisms in a lake or river can be affected by pollution. The statements below describe some of the effects on organisms of sewage being allowed to run into the water.

   The statements are in the wrong order.

Statement	Letter
Plants on the bottom of the lake die	A
Algae near the surface of the lake absorb nitrates and grow in large numbers	B
Sewage flows into the lake from a nearby farm	C
Increased algae prevent light from reaching plants rooted at the bottom of the lake	D
Bacteria break down the sewage into nitrates	E

   a. Rearrange the letters to show the correct sequence of events. The first one has been done for you.

Sequence of events	First step	Second step	Third step	Fourth step	Fifth step
Letter	C				

   [4]

   b. Suggest one other source of nitrates which may enter the lake.

   .... [1]

   c. The sequence that you have described in your answer to part **a.** is not the end of the harmful events that may take place.

   Describe three further steps to show how aerobic bacteria may cause the death of fish and larger invertebrates in the lake.

   1. ....

   2. ....

   3. .... [3]

Extension

2. One of the effects of eutrophication is that polluted water can be harmful to human babies. Suggest why this could happen.

Human impact on ecosystems

4.20 Conservation of species

1. Define the term *conservation*. ....
   .... [2]

2. a. The Giant Panda (*Ailuropoda melanoleuca*) is an endangered species.

   i. Suggest and explain **two** reasons why the panda is endangered. One reason should relate to the environment, and one reason should relate to the biology of the panda itself.

   Environmental factor .... Explanation of effect ....
   .... [2]

   Biological factor .... Explanation of effect ....
   .... [2]

   ii. The Giant Panda is sometimes called a **flagship species**. Explain what is meant by a flagship species. ....
   ....
   .... [2]

   b. A conservation management plan involves several steps. The first step is to sample the population of the endangered species.

   i. Explain why sampling is necessary. ....
   .... [2]

   ii. For any named species, suggest a suitable method of sampling.

   Name of species ....

   Method of sampling ....
   .... [2]

   c. Recent population surveys suggest that the hedgehog (*Erinaceus europaeus*) is very quickly declining in numbers. Some conservationists suggest that the badger (*Meles meles*) is to blame. Suggest what can be done to reverse this fall in hedgehog populations.

Extension

Human impact on ecosystems

4.21 Science and the fishing industry

1. Modern fishing boats use ultrasound equipment to locate shoals of fish, and giant nets with very small mesh to capture their prey.

   Describe and explain the possible harmful effects on organisms and the environment.

   ....
   ....
   ....
   ....
   ....
   .... [3]

2. Until recent times, fishing in the North Sea was largely unrestricted.

   Suggest **three** measures that could be taken to help to conserve the cod population in the North Sea.

   1. ....
   ....
   ....

   2. ....
   ....
   ....

   3. ....
   ....
   .... [3]

Extension

3. The Chinese government is building artificial islands on coral atolls in the South China Sea. Conservationists are anxious that this work will affect fish populations in that part of the world. Explain how this could happen.

Human impact on ecosystems

4.22 Worldwide conservation

1. A farmer believes in **sustainable** farming.

   a. What is meant by the term **sustainable**? ....

   .... [1]

   The farmer plants a forest of young fir trees. Ten years later the forest is thinned by removing some of the trees. A small part of the forest is harvested in each of the following years.

   b. Suggest **two** advantages of thinning the forest.

   ....

   .... [2]

   c. Suggest two disadvantages of growing trees as a monoculture.

   ....

   .... [2]

   d. This list contains some properties of trees. A farmer buying young trees would consider these properties when making his purchase.

fast growth	**good growth on poor soil**	**high quality wood for pulp**
	low cost	**resistance to disease**

   Choose any **three** of these properties and suggest a reason why each property would be useful to the farmer.

Property	Reason for choice

Extension

2. Explain why some users of oriental medicine believe that it is acceptable to **a.** kill rhinoceros and **b.** keep bears in small cages.

Human impact on ecosystems

4.23 Sewage treatment

1. This diagram shows one type of sewage treatment works.

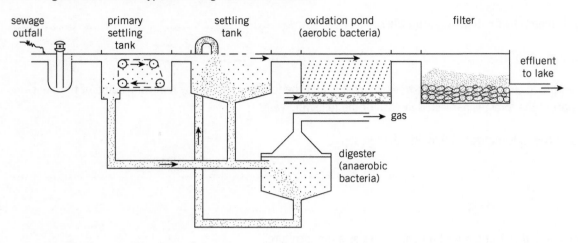

   a. Define the term *anaerobic*.

   .... [1]

   b. i. Name **one** gas produced by the microorganisms in the digester. ....

   .... [1]

   ii. Suggest **one** possible use of this gas.

   .... [1]

   c. Explain why it is important to treat sewage before it is released into the lake.

   ....

   ....

   .... [2]

   d. Processed sludge from the sewage treatment works can be used in agriculture. Describe **one** important agricultural use for the processed sludge.

   ....

   .... [2]

   e. Leakage of disinfectant into the sewers can be dangerous. Explain why disinfectants in sewers can cause harm.

   ....

   ....

   .... [2]

Extension

2. Some parts of the world are very prone to flooding. Explain how flooding can be a danger to the provision of safe drinking water.

Human impact on ecosystems

4.24 Saving fossil fuels

1. Biotechnology has made it possible to use algae as a source of fuel.

   The algae are grown in a vessel called a biocoil, then dried and ground up to form a powder, which can be used as fuel.

   The biocoil is a transparent tube about 5 m high, and the algae constantly circulate through it in a nutrient solution. The algae grow and multiply at a very fast rate.

   The diagram shows the main stages in this process.

   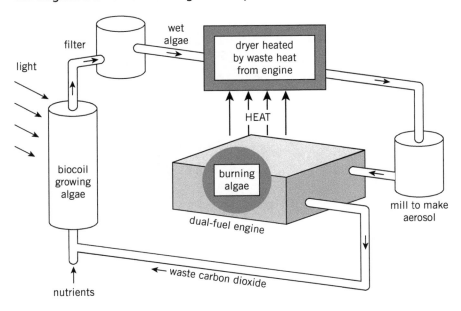

   a. Write out an equation to summarise how energy is trapped by the algae. [2]

   b. Explain why it is essential that the biocoil is transparent.

   ....

   ....

   .... [2]

   c. Explain why it is essential to use an organism with a high rate of reproduction in this process.

   ....

   ....

   .... [2]

Extension

2. Explain what is meant by 'fracking'.

   Suggest two possible benefits of this process, and two reasons why we might be anxious about the development of this industry.

Human impact on ecosystems

4.25 Paper recycling

1. a. The diagram shows some of the stages in paper manufacture and recycling.

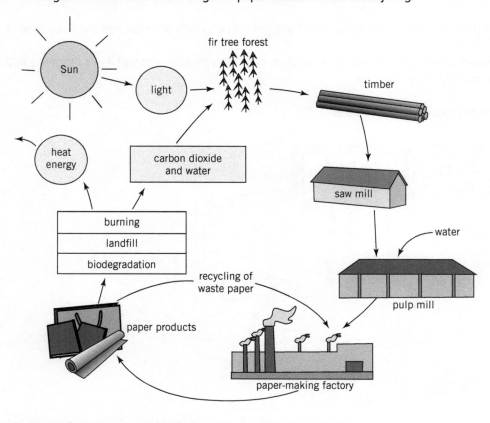

   i. Name the process used by the trees to manufacture sugars.

   .... [1]

   ii. Name the process by which carbon dioxide leaves the trees.

   .... [1]

   iii. In the mill the wood pulp is treated with sodium hydroxide to break down the fibres in the wood.

   State the effect on the pH of the pulp. .... [1]

   b. i. One of the paper products manufactured in the paper-making factory is paper bags. Give **two** reasons why paper bags are more environmentally friendly than plastic bags.

   ....

   .... [2]

   ii. Suggest **two** other uses of recycled waste paper.

   ....

   .... [2]

Extension

2. Assume that a soft drink can is composed of 100% aluminium, and that the mass of an Airbus-380 is 50% aluminium. Calculate how many soft drink cans would be needed to produce one Airbus-380. Consider where the aluminium would come from if it wasn't recycled.

Practical biology

5.1 Making a model of DNA

Although it took Watson and Crick many years of work, you can create a DNA model quite easily! You will be using some of the same information that Watson and Crick had available to them.

- DNA is made up of simple subunits called **nucleotides**.
- There are only four different nucleotides found in DNA – **adenine** (A), **guanine** (G), **cytosine** (C), and **thymine** (T). There is always the same amount of adenine as thymine, and the same amount of guanine as cytosine.
- There are **base pairing** rules: A always pairs with T, and G always pairs with C.

Alright, you are ready to go!

1. Cut out the block of nucleotides below, and separate each nucleotide from the others.

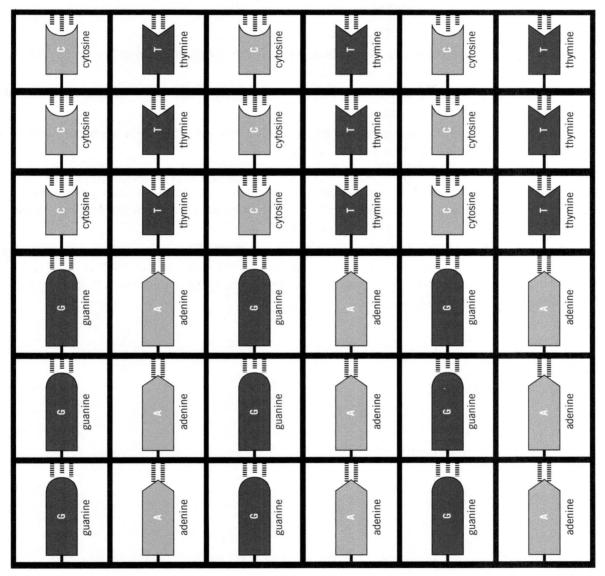

2. Place one of each of the nucleotides in the spaces below. Arrange them so that the 'bonds' hold the nucleotides in the correct base pairs.

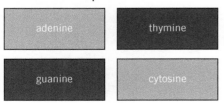

113

Practical biology
5.1 Making a model of DNA

   a. Use the diagram to explain why A always pairs with T, and G always pairs with C.

   b. What do you notice about the two 'letters' in each of the pairs?

3. Now create one strand of the DNA molecule. Do this by placing the correct cut-out nucleotide in the labelled spaces of the outline below.

4. Build up the 'matching' strand of DNA by using the base pairing rule (scientists use the word 'complementary' for 'matching'). Remember that the nucleotides in the complementary strand will be upside down.

CONGRATULATIONS! You have now produced a model of a short chain of DNA. If you are working as part of a large class group, perhaps some people in the class could build slightly different sequences, e.g.

AGGGCTTAAT   CCAGGCCTTA   TAGCCAGTAG

GGCATTGCCC   CGCGATATTG   AGCCCTTAGG

If possible, you should now make a copy of your piece of DNA. If your teacher has given you an outline for the DNA model, you can stick the nucleotide 'letters' into their correct positions.

Now you could try something really tricky! In 'real' DNA, the two strands are coiled around each other to make a **double helix**. The twisting takes place once every ten pairs, so....

You can now twist your piece of DNA so that there is just one turn in it. You should be able to clearly see the top base pair and the bottom base pair. Now you have a piece of **double helix**.

A gene is a section of DNA that codes for one protein in a cell. Even a small protein needs quite a long gene. If you join up your piece of DNA with pieces from your classmates, you might get a good idea of a single gene. For example, about 30 pieces joined together in a long double helix would give enough coded information to make a simple protein like **insulin** (the protein hormone which controls our blood sugar level).

A human cell contains more than 10 000 genes in its nucleus. You need so much DNA to make up these genes that each of your cells (which you can't even see without a microscope!) contains about 2 metres of DNA. The DNA has to coil up like balls of wool to fit in! A mature human has so many cells that there is enough DNA in each of us to stretch to the Moon............and back...........and to the Moon again!

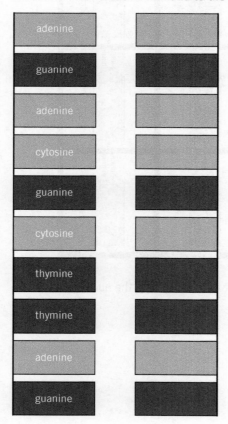

Practical biology
5.2 Drawing skills: the structure of flowers

A flower is made up of a set of modified leaves. The leaves are arranged in rings which produce the gametes, protect them, and ensure that fertilisation takes place. The rings of leaves are attached to the end of the flower stalk, the receptacle.

Examination of a typical flower

If a typical flower (a buttercup, for example) is examined, it is possible to see the four rings of specialised leaves. It may be necessary to use a hand lens to do this easily.

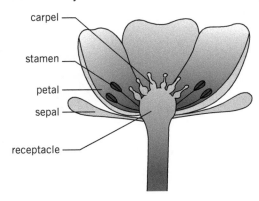

Arrange the four rings (carpel, sepal, stamen, and petal) from outside to inside.

Use a scalpel (take care!) and a pair of forceps to remove a representative part of each ring of leaves. Examine the structure under a hand lens or binocular microscope.

The individual structures will look something like those shown in this diagram.

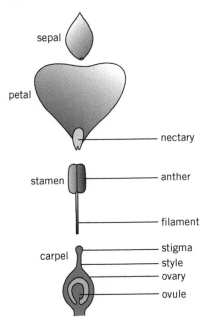

How many of each of the structures can you find in your flower?

Variations in flower structure

Flowers of different species may differ in:

- The number of each of the different components.
- The arrangement of the different parts – especially how fused (stuck together) they are to form tubes or platforms.

Practical biology

5.2 Drawing skills: the structure of flowers

Drawing a flower

With so much variation in flower structure, scientists have developed standard ways of describing flower structure. The most common method is to draw a **half flower**.

Use a razor blade or scalpel to cut through the flower, down the line of its stalk. Hold the flower in forceps if necessary, and take care with the sharp blades.

Now draw the flower in section (this gives a three-dimensional view of the petals and sepals).

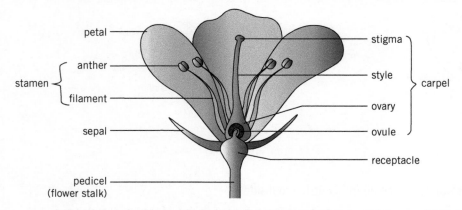

Make sure that the structures are labelled accurately with guide lines drawn with a ruler. (Guide lines always look neater when they are parallel to the top and bottom of the page!)

Add a scale to your drawing

[6]

Practical biology

5.3 Germination (follow on from 3.5)

You are asked to design an experiment to study the effect of temperature on the germination success rate of pea seeds. The teacher suggests a temperature range of 5–55 °C.

a. State the independent, the dependent, and any fixed variables that you should include in your experimental design.

Independent variable	Dependent variable	Fixed variables

[5]

b. Suggest how you would measure the value of the dependent variable. [1]

c. Draw a table in which you could present your results.

[3]

d. Prepare a grid on which the results could be plotted. Draw a graph of the likely results of your investigation.

[4]

e. Explain the results which you have presented in the graph.

....
....
....
.... [3]

117

Practical biology — 5.4 Transpiration experiment

The diagram shows a piece of apparatus that can be used to measure the rate of water uptake by a plant shoot.

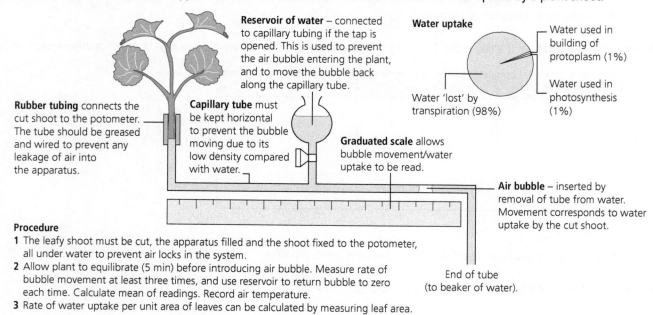

Procedure
1 The leafy shoot must be cut, the apparatus filled and the shoot fixed to the potometer, all under water to prevent air locks in the system.
2 Allow plant to equilibrate (5 min) before introducing air bubble. Measure rate of bubble movement at least three times, and use reservoir to return bubble to zero each time. Calculate mean of readings. Record air temperature.
3 Rate of water uptake per unit area of leaves can be calculated by measuring leaf area.

Two sets of this apparatus were used at the same time. The two sets of apparatus had shoots taken from the same tree. The experiment was carried out with four different sets of external conditions **A**, **B**, **C**, and **D**. The time for the air bubble to move 10 cm was measured and recorded in the table.

	External conditions	Time for air bubble to move 10 cm / s	
		shoot 1	shoot 2
A	dry, still air at 15 °C	25	46
B	dry, still air at 25 °C	19	37
C	dry, moving air at 25 °C	16	32
D	humid, still air at 15 °C	58	78

a. State which shoot took up water most quickly under all conditions.

.... [1]

b. Suggest a difference between the shoots that could explain these results.

....

.... [1]

c. Explain why the results were different under condition **D** than from the condition in **A**.

....

....

.... [2]

Practical biology — 5.5 Variables

A student carrying out food tests for biological molecules had seen on the internet that **cow's milk contained more sugar than soya milk**. He suggested that the group should test this hypothesis.

a. Describe how you would carry out the test for a simple sugar.

....

....

.... [3]

b. The test would only be valuable if it is **quantitative**. State the meaning of the term quantitative.

.... [1]

c. The teacher suggested that the students should fill in this table, to make sure that they had a plan that would provide them with valid data. Complete the table.

Which variable would you change in the experiment? This is the **independent variable**.	
Which three variables would you keep the same? These are the **fixed variables**.	
Which variable would you measure to test the hypothesis? This is the **dependent variable**.	
How would you measure this variable?	
Suggest a control for the experiment.	
Suggest why you would repeat the experiment.	

[8]

Mathematics for biology

6.1 Measurement and magnification

1. The **size** of a structure or organism is measured in units of **length** (such as mm or m). When a diagram is made, or a photograph taken, it may not be easy to directly show the correct size – for example, when a structure is extremely small or very large.

   The correct (or true) size of an organism can be calculated using a combination of actual measurement and a known magnification.

   There are two simple relationships that should be understood:

   Magnification $= \dfrac{\text{measured length}}{\text{actual length}}$

   Actual (true) length $= \dfrac{\text{measured length}}{\text{magnification}}$

   It is also important that candidates can use a **scale line** to work out magnification.

   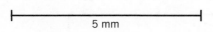

   5 mm

   This means that the line drawn represents 5 mm in actual length.

   So, magnification $= \dfrac{\text{measured length of scale line}}{\text{actual length of scale line}}$

   $= \dfrac{85}{5} = 17$

   Note that there are no units for magnification – it is a comparison of lengths. Be careful to make sure that the two lengths you are comparing are given in the same units.

   State

   i. How many mm there are in 1 cm ....

   ii. How many μm there are in 1 mm..

   Core students (Papers 1 and 3) can use millimetres (mm) as units, but candidates taking extension papers (Papers 2 and 4) should also be confident with the use of **micrometres** (μ or μm).

   The diagram shows a cell from the pancreas of a human.
   The cell was drawn using a light microscope.

   a. Identify the structures labelled **A**, **B**, and **C**. [3]

   b. Which of these structures would not be present in a red blood cell? [1]

   c. Name two additional structures that you would see in a palisade cell from a leaf. [2]

   d. Use the scale shown alongside the cell to calculate how much it has been magnified. Show your working. [3]

   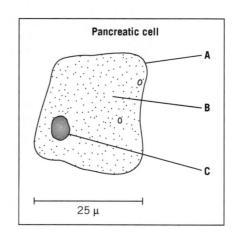

Mathematics for biology

6.2 Multiple births

1. Approximately 1% of all births are **multiple births**, i.e. more than one baby develops and is born at the same time.

   This table shows the frequency of multiple births in mothers in different age groups.

Age group of mother (years)	Frequency of multiple births (percentage of all live births)
Younger than 20	0.6
20–24	0.7
25–29	1.0
30–34	1.2
35–39	1.5
40–44	1.2
Older than 45	1.4

   a. Plot this information as a histogram. Use the grid provided below.

   b. Comment on the effect of the mother's age on the frequency of multiple births.

   ....

   ....

   .... [2]

121

Mathematics for biology

6.3 Enzyme experiments

1. The rate of activity of amylase is affected by temperature. The effect of temperature was investigated by using a set of six identical test tubes containing 5 cm³ of starch solution. Each test tube was placed in a water bath at a different temperature and then 1 cm³ of amylase solution was added to start the reaction.

   The time taken for the starch to disappear was measured. The results are recorded in this table.

Temperature (°C)	20	25	30	35	40	45
Time taken for starch to disappear (s)	600	315	210	175	200	420

   a. Use the grid provided to plot a graph of these results.

   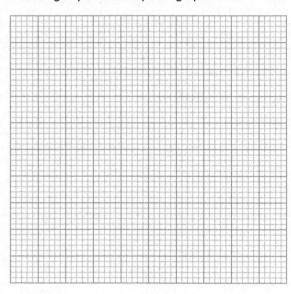

   [4]

   b. State the temperature at which the amylase works best.

   ....°C [1]

   c. Explain why it is important that the same volume of starch and amylase was present at the start of the experiment.

   ....

   ....

   .... [1]

   d. Chemical reactions usually get faster as temperature increases. Suggest a reason why the rate of amylase activity does not increase above 40 °C.

   ....

   ....

   .... [2]

   e. Name one other factor that would affect the rate of amylase activity.

   .... [1]

Mathematics for biology

6.4 The control of photosynthesis

1. Scientists were able to grow red peppers and lettuces in large greenhouses. This table shows the effects on crop yields of adding extra carbon dioxide to the atmosphere inside the greenhouses.

Crop	Crop yield (kg)	
	Normal air (0.04% CO_2)	Enriched air (0.08% CO_2)
Red peppers	0.42	0.63
Lettuces	0.9	1.1

   a. Calculate the percentage increase in yield for red peppers when the air is enriched with carbon dioxide. Show your working.

   .... [2]

   b. There are a number of possibilities for increasing the carbon dioxide concentration in the greenhouse. Study this list, then choose the best method. Explain your decision.

   - Relying on the exhaled carbon dioxide from greenhouse workers.
   - Adding extra manure to the soil so that microbes can respire and release carbon dioxide.
   - Releasing carbon dioxide from cylinders of compressed gas.
   - Using paraffin heaters in the greenhouse, as burning paraffin releases carbon dioxide.

   Best method ....

   Explanation ....

   .... [2]

Mathematics for biology

6.5 Ingestion

1. The pH of the saliva of 100 students in a school was measured and compared with the number of teeth which had required fillings.

   The results are presented in this table.

pH of saliva	Percentage of pupils in each pH range	Average number of fillings per student in each pH range
6.7–6.9	20	8
7.0–7.2	60	5
7.3–7.5	20	3

   a. Suggest how the pH of the saliva could be measured.

   ....

   .... [2]

   b. State the hypothesis being tested in this investigation.

   .... [1]

   c. State and explain whether the results support the hypothesis.

   ....

   ....

   .... [2]

Mathematics for biology

6.6 The leaf and water loss

1. An experiment was carried out to compare the water loss from a leaf and from a piece of damp blotting paper. The leaf and paper were hung on a string as shown in this diagram.

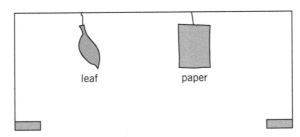

The mass of each was recorded every hour for a period of five hours. The results are shown in this table.

Time (hours)	Mass (g)	
	Paper	Leaf
0	6.1	6.5
1	5.4	6.1
2	4.7	5.7
3	4.0	5.8
4	3.3	5.9
5	2.6	4.6

a. Plot these results on the grid below.

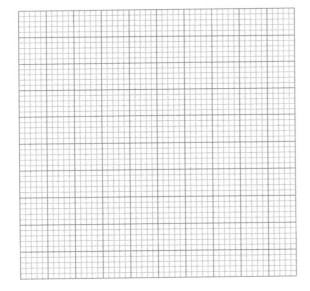

[5]

b. Explain why it was important that the leaf and the paper were the same size and shape.

....

....

....

....

.... [2]

c. i. State the mass of the paper after 2.5 hours. .... g [1]

ii. Calculate the rate of loss of mass by the leaf during the first two hours of the experiment. Show your working.

[2]

125

Mathematics for biology

6.7 Individuals and the community

1. This table shows the causes of death amongst cigarette smokers in the UK.

Cause of death	Percentage of deaths
Lung cancer	
Bronchitis and emphysema	7
Diseases of the circulatory system	10
Other causes (not directly related to smoking)	75

   a. Complete the table to show the percentage of smokers who die from lung cancer. [1]

   b. Use the data in this table to draw a pie chart.

   Draw this in the circle below. Label the pie chart.

   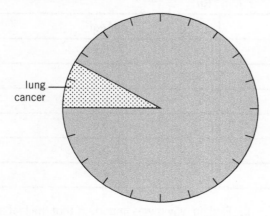

   [3]

   c. Suggest how the community might be able to reduce the number of smoking-related deaths.

   ....

   ....

   .... [2]

Mathematics for biology

6.8 The measurement of respiration

1. A group of students wanted to compare the rate of respiration in four different fruits, as well as in tomatoes.

   a. For this experiment, state the **independent variable** and the **dependent variable**, and suggest one variable that should be **fixed**.

Independent variable	Dependent variable	Fixed variable

   [3]

   b. This table contains their results.

Tomato	1.55	1.50	1.45	1.50
Damson	0.75	0.75	0.75	0.75
Apple	1.00	1.05	0.95	1.00
Strawberry	1.26	1.20	1.23	1.23
Blackberry	0.90	0.90	0.90	0.90

   The students forgot to provide headings for the columns in their table.

   Suggest suitable column headings and write them into the spaces at the head of each column.

   [4]

   c. Use the grid below to plot a bar chart of their results.

   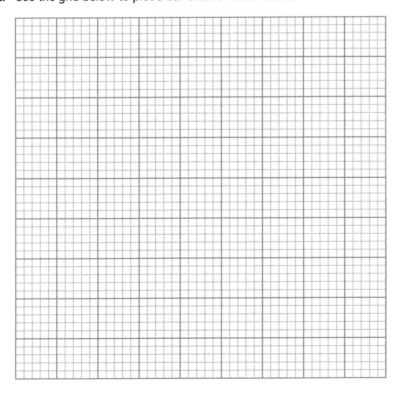

   [4]

127

Mathematics for biology

6.9 Formation of seed and fruit

1. A broad bean fruit (a pod) contains several seeds. A student opened up 50 broad bean pods and counted the number of seeds in each. He recorded his results in a tally chart.

Number of seeds in pod	4	5	6	7	8
Tally	II	⊪⊪	⊪⊪⊪⊪ II	⊪⊪⊪	I
Total number					

   a. Complete the table to show the number of pods containing each number of seeds. [1]

   b. Present these results in a suitable graph. Use the grid provided below.

   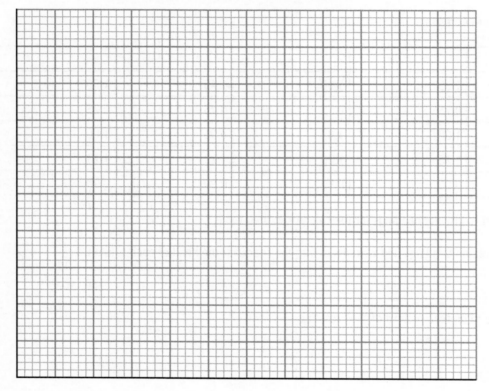

   [4]

   c. State the most frequent number of seeds in a pod. .... [1]

   d. Calculate the percentage of pods which contain the most common number of seeds. Show your working.

   ....% [2]

128

Mathematics for biology

6.10 Variation and selection

1. Jack and Salim were working together on a mathematical investigation. They decided to measure the body masses of each of the other students in their class.

    The results are shown in this table.

Mass category (kg)	Tally of number in group	Number in group
35–38	I	
39–42	II	
43–46	IIII	
47–50	IIII I	
51–54	IIII IIII	
55–58	IIII	
59–62	III	
63–66	II	
67–70	I	

    a. Complete the third column of the table by adding up the tally numbers. [2]

    b. Plot these results in a suitable form on the grid below.

    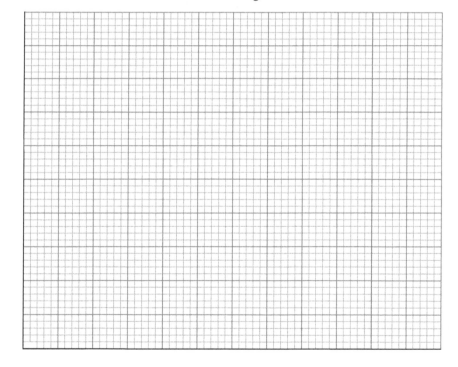

    [5]

    c. State which form of variation this graph illustrates. .... [1]

    d. Explain how this form of variation comes about. ....

    .... [2]

Mathematics for biology

6.11 The size of populations

1. An experiment was carried out to follow the population growth of the bacterium *Escherichia coli*.

   Cells of the bacterium were placed in a flask containing sterile nutrient broth and then incubated at 30 °C. Every five hours a sample was removed from the flask, and the number of bacteria present was calculated.

   The results are shown in this table.

Time from start (h)	5	10	15	20	25	30	35	40	45	50
Number of bacteria (millions per cm³)	20	50	430	450	460	420	220	55	20	0

   a. Draw a graph of these results on the grid below.

   b. Suggest why the nutrient broth:

   contains sugar .... [1]

   contains amino acids .... [1]

   must be sterile. ....

   .... [2]

   c. Mark with a letter **R** on the graph the period during the experiment when the rate of reproduction was at its greatest.
   [1]

   d. Suggest **two** possible reasons why the population declined between 25 and 50 hours. [1]

   ....

   .... [2]

Mathematics for biology

6.12 Human population changes

1. The graph below shows the change in world population since 1600.

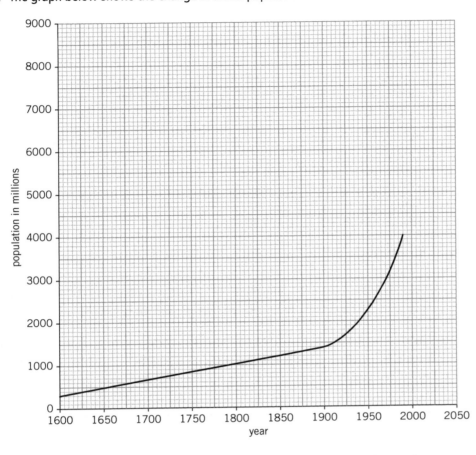

a. Describe the **two** main changes in population growth shown in the graph. [1]

1. ....

2. ....
.... [2]

b. State the year in which the world population reached 2.5 billion.
.... [1]

c. State the world population in 1850.
.... [1]

d. Calculate how long it took for the world population to double in size after the year 1800. Show your working.

.... [2]

e. Use the curve to estimate the likely world population in 2050.
.... [1]

131

Revision
7.1 Analysing command words

Exam success: knowing what to do

Candidates taking an examination in IGCSE Biology are given instructions about what the examiner expects from them. These instructions are given in the introduction to each question, or to each part of a multi-part question. These instructions, which tell the candidate what to do, are given as **command words**.

For example, a candidate might be asked to **define** a biological term, to **describe** a biological process, or to **calculate** a numerical value. Define, describe, and calculate are command words. To be successful in an examination, a candidate must understand what each of the command words means.

Command words: what answer do you expect?

- Command words may require either concise answers or extended answers.
- Command words may require either recall or making logical connections between pieces of information.
- Some command words require only single word or single figure answers.

This list of the command words is taken from the IGCSE syllabus, published by CIE. The examination board stresses that these words are best understood when they are seen in an actual question, and they also point out that some other command words may be used. Even so, this list contains the most commonly used words and what they mean. In other words, these command words tell you what the examiner wants you to do when trying to answer a particular question.

Name: the answer is usually a technical term (diffusion, for example, or mitochondrion) consisting of no more than a few words. **State** is a very similar command word, although the answer may be longer, as a phrase or sentence. **Name** and **state** don't need anything added, i.e. *there is no need for an explanation*. Adding an explanation will take up time and probably won't gain any more marks!

> a. State **three** normal functions of a root.
>
> 1. To anchor the plant in the soil
> 2. To absorb water from the soil
> 3. To absorb mineral ions from the soil
>
> [3]

Define: the answer is a formal definition of a particular term. The answer is usually 'what is it' – for example, define the term active transport means 'what is active transport'.

Look carefully how many marks are offered. It is often a good idea to add an example to a definition – in this way the examiner can be sure that you know what you are trying to define.

What do you understand by or **what is meant by** are commands that also ask for a definition, but again the marks offered suggest that you should add some relevant comment on the importance of the terms concerned.

List: you need to write down a number of points, usually of only one word, with no need for explanation. For example, you might be asked to list four characteristics of living organisms.

Revision
7.1 Analysing command words

Describe: your answer should simply say what is happening in a situation shown in the question, e.g. the number of germinating seeds increased to 55. There is no need for an explanation.

No call for explanation or comment

Make **three** points for the **three** marks on offer

(c) Describe how water reaches a leaf and enters a palisade cell.

Evaporation/transpiration from leaf surface

water pulled up xylem to replace 'losses'

water moves by osmosis to palisade cell

water crosses palisade cell membrane by osmosis

[3]

(d) Describe how sugar produced in a palisade cell reaches the roots.

Conversion to sucrose/transported in phloem/

unloaded in roots by diffusion or active transport

[2]

Explain: the answer will be in extended prose, i.e. in the form of complete sentences. You will need to use your knowledge and understanding of biological topics to write more about a statement that has been made in the question, or earlier in your answer.

The command word **explain** is often linked with **describe** or **state**, so that the examiner asks you to **describe and explain** or **state and explain**. This means that there are two parts to be answered – they can often be joined by the word 'because'. For example 'the number of germinating seeds increased to 55 (describe) **because** the temperature has increased and germination is controlled by enzymes which are sensitive to temperature (explain)'.

Many answers fail to gain full marks because they do not obey both commands – they often **describe** but do not **explain**, so only gain half of the available marks.

---

c. There is concern that pollution of the environment may change the breeding grounds of the Adelie penguin.

State and explain the effect this might have on the populations of the Leopard seal and the Ross seal.

two pieces of information needed – can you link them with the word 'because'?

Leopard seal Population could fall because there would be less food for them (so they would breed less successfully)

Ross seal Population could rise because there would be fewer leopard seals to act as predators

[4]

133

Revision

7.1 Analysing command words

Suggest: this command word has two possible meanings. In the first, you will need to use your biological knowledge and understanding to explain something that is new to you. You might use the principle of enzyme action (which you know about) to explain an industrial process (which might be new to you). Suggest also has a second meaning – it implies that that there may be more than one possible answer to a question. For example, there might be a number of different factors affecting the action of a digestive enzyme.

Questions which involve data response (like looking at a table of results, for example) and problem solving (like comparing three different environmental situations, for example) often begin with the command word **suggest**.

Calculate: a numerical answer is expected, usually obtained from data given in the question. Remember:

- Show your working (there may be marks given for the correct method even if you get the wrong answer).
- Give your answer to the correct number of significant figures, usually two or three.
- Give the correct units (if needed – sometimes these will already be given in the space left for your answer).

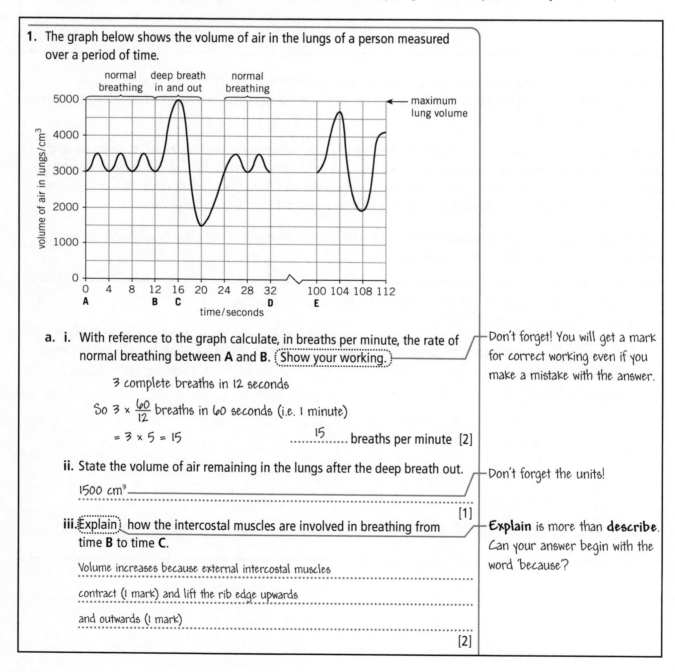

1. The graph below shows the volume of air in the lungs of a person measured over a period of time.

   a. i. With reference to the graph calculate, in breaths per minute, the rate of normal breathing between **A** and **B**. Show your working.

   3 complete breaths in 12 seconds

   So $3 \times \frac{60}{12}$ breaths in 60 seconds (i.e. 1 minute)

   $= 3 \times 5 = 15$

   ....15.......... breaths per minute [2]

   — Don't forget! You will get a mark for correct working even if you make a mistake with the answer.

   ii. State the volume of air remaining in the lungs after the deep breath out.

   1500 cm³

   — Don't forget the units!

   [1]

   iii. **Explain** how the intercostal muscles are involved in breathing from time **B** to time **C**.

   Volume increases because external intercostal muscles

   contract (1 mark) and lift the rib edge upwards

   and outwards (1 mark)

   — **Explain** is more than **describe**. Can your answer begin with the word 'because'?

   [2]

134

Revision

7.1 Analysing command words

Other terms which require numerical answers are **find**, **measure**, and **determine**.

Find is a general term, and can mean calculate, measure, or determine.

Measure implies that the answer can be obtained by a direct measurement, e.g. using a ruler to measure the length of a structure on a diagram.

Determine means that the quantity cannot be measured directly, but has to be obtained by calculation or from a graph. For example, the size of a cell from a scale included in the diagram of the cell.

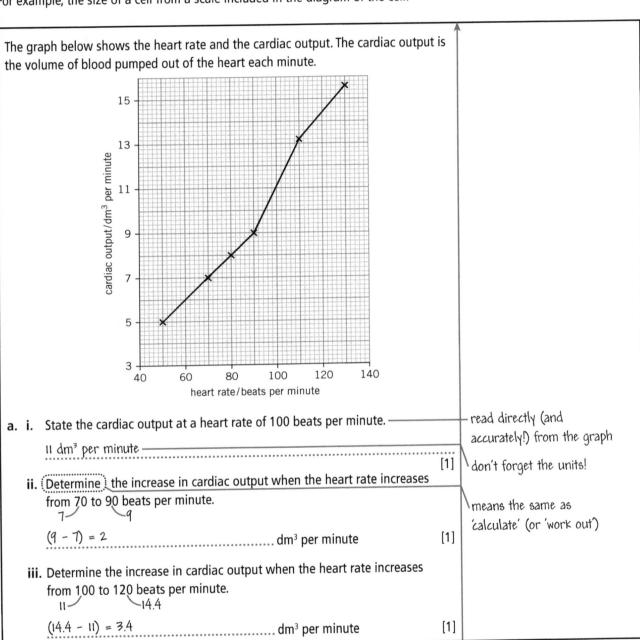

The graph below shows the heart rate and the cardiac output. The cardiac output is the volume of blood pumped out of the heart each minute.

a. i. State the cardiac output at a heart rate of 100 beats per minute. — read directly (and accurately!) from the graph

   11 dm³ per minute

   [1] — don't forget the units!

   ii. Determine the increase in cardiac output when the heart rate increases from 70 to 90 beats per minute.
   7 — 9

   (9 − 7) = 2 .... dm³ per minute [1]

   — means the same as 'calculate' (or 'work out')

   iii. Determine the increase in cardiac output when the heart rate increases from 100 to 120 beats per minute.
   11 — 14.4

   (14.4 − 11) = 3.4 .... dm³ per minute [1]

Revision

7.1 Analysing command words

- Some questions cover complicated material but actually contain most of the information that you need to gain full marks!

7. a. Restriction endonucleases are enzymes used to cut DNA into fragments during genetic engineering. EcoR1 is a restriction enzyme produced by the bacterium *Escherichia coli*. It always cuts DNA at the same sequence of organic bases. The base sequence below shows some of the remaining organic bases after a cut.

   Write the missing bases in the correct position to show the DNA before the cut. — many candidates missed this instruction (and hence 1 mark)

   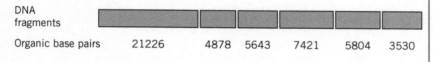

   [1] — could be easily worked out as long as you noted the question 'gave' the pairing of G–C, the other base pair therefore **must** be A–T (or T–A)

   b. The diagram below represents a length of DNA after being treated with EcoR1 restriction endonucleases.

   DNA fragments

   Organic base pairs   21226   4878   5643   7421   5804   3530

   i. State how many   C T T A A G   base sequences were on the original DNA molecule.

   ....5........ [1] — i.e. there were five points at which this enzyme could recognise and 'cut' the DNA

   ii. Work out how many individual organic bases there are on the DNA shown.

   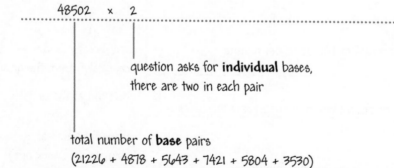

   [1]

   question asks for **individual** bases, there are two in each pair

   total number of **base** pairs
   (21226 + 4878 + 5643 + 7421 + 5804 + 3530)

Revision

7.1 Analysing command words

c. The following table lists events from the identification of a human gene coding for a hormone "X" to the commercial production of hormone "X".

Genes can be transferred into plasmids, tiny circles of DNA which are found in bacteria.

Show the correct sequence of events 1 to 8 by writing the appropriate number in each box provided. The first (number 1) and last (number 8) have been completed for you.

Event	Number
Cutting of a bacterial plasmid using restriction endonuclease	3
Cutting of human DNA with restriction endonuclease	2
Identification of the human DNA which codes for hormone "X"	1
Many identical plasmids, complete with human gene, are produced inside the bacterium	6
Mixing together human gene and "cut" plasmids to splice the human gene into the plasmid	4
Some of the cloned bacteria are put into an industrial fermenter where they breed and secrete the hormone	8
The bacterium is cloned	7
Using the plasmid as a vector, inserting it, complete with human gene, into a bacterium	5

[6]

Annotations:
- *follow very straightforwardly from number 1.* (pointing to rows 1 & 2)
- *clearly the step prior to number 8* (pointing to "The bacterium is cloned")
- *Six marks gained by applying a logical and methodical approach to dealing with information supplied by the examiner.*

Simple summary:

State, define, or **name**	What is it?
Describe	What is happening?
Explain	Why is it happening?
Calculate	How many or how big?

Check a copy of a recent IGCSE Biology Paper (0610 for example). Underline the command words. Which ones are used most often?

Exam-style questions

8.1 Exam-style questions

Multiple choice

Remember that there are two alternative multiple choice papers. Paper 1 contains questions which only assess material from the **core** syllabus, paper 2 contains questions that could contain material from either the **core** or the **supplement**.

Sample 'core' questions

1. Root hair cells are found on plant roots.

   Which of the following features would be present in a root hair cell but not in a cell lining the small intestine?

   A   cytoplasm   B   chloroplasts

   C   cell wall   D   cell membrane

2. Which animal is a spider?

   1. has legs .... go to 2

      has no legs .... go to 3

   2. has six legs .... organism **A**

      has eight legs .... organism **B**

   3. has a shell .... organism **C**

      has no shell .... organism **D**

3. Dietary fibre passes through several structures after leaving the stomach.

   In which order does the dietary fibre pass through these structures?

   A   ileum – duodenum – rectum – colon

   B   duodenum – ileum – rectum – colon

   C   ileum – duodenum – colon – rectum

   D   duodenum – ileum – colon – rectum

4. Which chemical elements are found in carbohydrates, fats, and proteins?

	Carbohydrates	Fats	Proteins
A	Carbon, hydrogen, and oxygen	Carbon, hydrogen, and oxygen	Carbon, hydrogen, oxygen, and nitrogen
B	Carbon, hydrogen, and oxygen	Carbon, hydrogen, oxygen, and nitrogen	Carbon, hydrogen, and oxygen
C	Carbon, hydrogen, oxygen, and nitrogen	Carbon, hydrogen, and oxygen	Carbon, hydrogen, and oxygen
D	Carbon, hydrogen, oxygen, and nitrogen	Carbon, hydrogen, and oxygen	Carbon, hydrogen, oxygen, and nitrogen

Exam-style questions
8.1 Exam-style questions

5. The diagram shows a germinated bean seed with a horizontal radicle. This is placed on a slowly rotating disc and is left for three days.

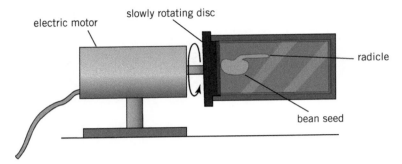

Which diagram shows the appearance of the radicle after three days?

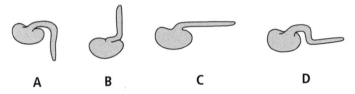

6. The diagram shows a section through the eye.

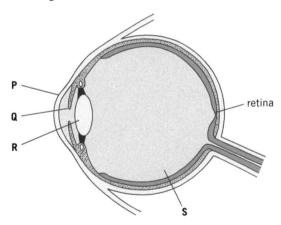

Which pair of structures focus light rays onto the retina?

A P and Q   B P and R   C Q and R   D Q and S

Sample 'supplement' questions

7. The graph shows the rate of growth for a population of herbivores.

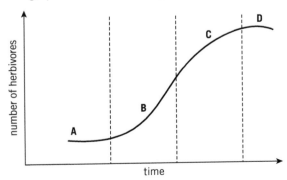

139

Which is the exponential (log) phase for the growth of the population?

8. Which graph shows the effect of temperature on the rate of photosynthesis?

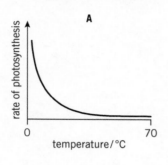
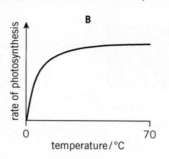
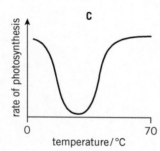
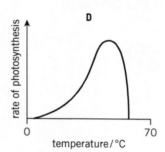

9. The diagram shows part of a transverse section of a leaf.

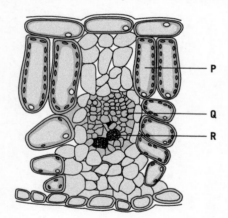

Which cells conduct water into the leaf and which cells conduct sugars out of the leaf?

	Conduct water	Conduct sugars
A	P	Q
B	Q	P
C	Q	R
D	R	Q

10. The diagram shows four places on a river. Water samples were taken at each place.

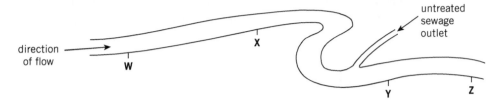

Which graph shows oxygen concentrations in the river?

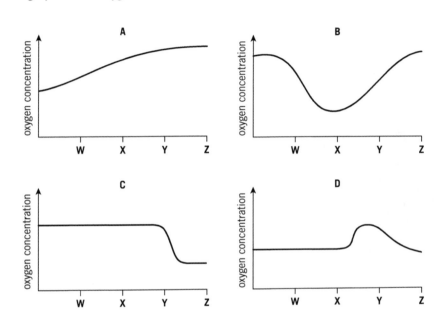

Exam-style questions

8.1 Exam-style questions

Longer written answers

Remember that there are also two alternative written papers. Paper 3 contains questions that only assess material from the **core** syllabus, paper 4 contains questions that could contain material from either the **core** or the **supplement**.

Sample 'core' questions

11. The diagram below shows the human alimentary canal.

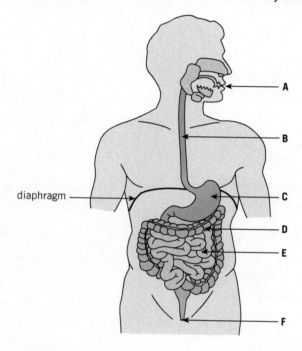

a. On the diagram, draw the liver in the correct place and approximately the right size. [2]

b. State the name of the process that moves food along the gut.

   .... [1]

c. Use the diagram to complete the following table.

   Each letter may be used once, more than once or not at all.

Description of site	Letter
Where saliva is released	
Where food is chewed	
Where hydrochloric acid is produced	
Where soluble foods are absorbed	
Where faeces are egested	

   [5]

142

12. The diagram below shows how carbon may be recycled.

    The letters represent processes that occur in the cycle.

    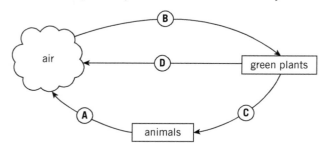

    a. i. Use words from the following list to identify the processes **A**, **B**, **C**, and **D**.

    feeding   **photosynthesis**   **respiration**

    The words may be used once, more than once, or not at all.

Letter	Process
A	
B	
C	
D	

    [4]

    ii. Name and describe **one** other process which is part of a complete carbon cycle.

    ....

    ....

    ....

    .... [3]

b. The diagram below shows a compost heap and the materials used to make it.

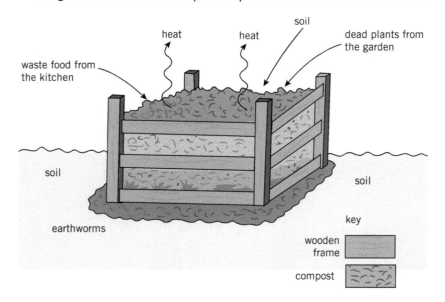

143

i. Explain why there is a difference in the temperature between the compost heap and the air surrounding it.

....

....

.... [2]

ii. Suggest one reason why a gardener uses an open frame to support the compost heap, rather than closed sides of the frame.

.... [1]

13. The diagram below shows a section through an air sac and a surrounding blood capillary.

   a. i. List three features shown in the diagram that make an air sac an efficient site for the exchange of gases.

   1. ....

   2. ....

   3. .... [3]

   ii. Suggest why the walls of the air sac contain elastic fibres.

   .... [1]

b. Use the bar graph to identify **two** differences between the composition of the blood at **A** compared with the composition of the blood at **B**.

A and **B** are shown in the diagram for part **a**.

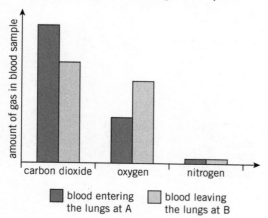

Exam-style questions   8.1 Exam-style questions

1. ....
   ....

2. ....
   .... [2]

c. Suggest **three** ways in which smoking can reduce the efficiency of the lungs.

1. ....
   ....

2. ....
   ....

3. ....
   .... [3]

14. Phenylketonuria is a disease affecting the central nervous system of humans.

    This disease is caused by a single gene mutation.

    a. i. Define the term *mutation*.

       ....
       .... [1]

       ii. State the name of **one** other disease affecting humans which is the result of a gene mutation.

       .... [1]

    b. The diagram shows part of a family tree showing phenylketonuria.

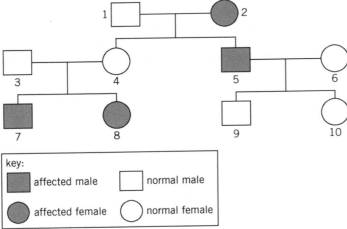

145

Exam-style questions
8.1 Exam-style questions

i. Phenyketonuria is caused by a recessive allele, **p**.

   Define the term *recessive*. ....

   .... [2]

ii. State the phenotype of individual 8. .... [1]

iii. State the possible genotypes of: individual 4 ....

   individual 6. .... [2]

iv. Individual 10 marries a man who is heterozygous for this condition.

   Use a Punnett square to find the probability that their first child will have phenylketonuria.

   Probability .... [4]

Sample 'supplement' questions

15. The diagram shows apparatus used to investigate the action of yeast on glucose solution.

    In Experiment **A**, 3 g of yeast were added to 10% glucose solution which had previously been boiled and cooled to 30 °C. The mixture was then placed into a vacuum flask.

    The procedure was repeated for Experiment **B** except the temperature was reduced to 20 °C.

    The time taken for the hydrogencarbonate indicator to turn from orange-red to yellow was measured and noted.

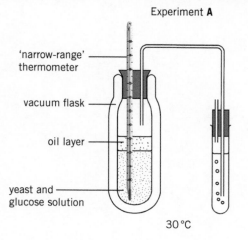
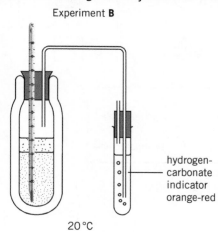

The results of the investigation are shown in this table.

	Experiment A	Experiment B
Temperature at start of experiment (°C)	30.0	20.0
Temperature 3 hours later (°C)	31.2	20.5
Time taken for indicator to change colour (min)	58	93

a. Explain why the time taken for the indicator to change colour was different in experiments **A** and **B**.

....

.... [2]

b. i. Suggest why a layer of oil was placed on the surface of the solutions in flasks **A** and **B**.

.... [1]

ii. Explain why the 10% glucose solution was boiled before it was cooled and mixed with the yeast.

.... [1]

c. Write a chemical equation to describe the living process which is occurring inside the two flasks.

[2]

16. a. Sickle cell anaemia is a condition caused by an abnormal form of haemoglobin.

The gene for production of haemoglobin exists in two alternative forms.

H_A codes for normal haemoglobin H_S codes for abnormal haemoglobin

i. State the name for the alternative forms of a gene. .... [1]

ii. A child has sickle cell anaemia. The parents do not have this disorder. Draw a genetic diagram to show how the child inherited the disorder.

[4]

iii. The parents are about to have another child.

State the probability that this child will have sickle cell anaemia. .... [1]

b. The maps below show the distribution of sickle cell anaemia and malaria in some parts of the world.

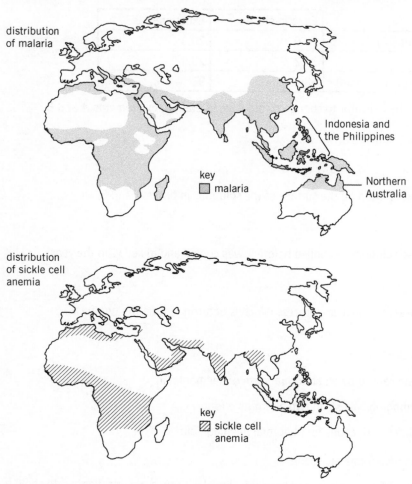

i. Explain why sickle cell anaemia is common in people who live in areas where malaria occurs.

....

....

.... [3]

ii. Suggest why sickle cell anaemia is very rare among people who live in Indonesia and Northern Australia.

....

.... [2]

Project ideas

9.1 Modelling neurones

ACTIVITY: MAKING MODEL NEURONES

You will need:

- Lengths (about 20–25 cm) of multi-strand single core wire, with plastic insulation. Ideally use wire with white or yellow insulation
- Small beads, with a hole for threading
- Wire strippers

Method:

1. **Motor neurone**

   - Strip 10 mm of plastic from the end of the first length of wire.
   - Push the exposed inner core through the thread hole in the bead. Push the bead up close to the plastic insulation, and then spread out the protruding strands from the inner core.
   - Strip 5 mm of plastic from the other end of the wire. Spread out the strands of the core wire that you have exposed.

2. **Sensory neurone**

   - Strip 25 mm of plastic insulation from one end of the wire. Thread on a bead as above, but only spread out the final 2 or 3 mm of inner core strands.
   - Strip 5 mm of plastic from the other end of the wire. Spread out the strands of the core wire that you have exposed.

3. Draw your two models. Use a library book or the internet to find diagrams of sensory and motor neurones so that you can label your models.

Extension

Why do you think it is important that the neurones are covered with an insulating layer?

In the disease **multiple sclerosis**, this insulating layer is not properly made. Find out the symptoms of this disease, and suggest how this loss of insulation is responsible for the symptoms.

149

Project ideas

9.2 Modelling the spinal cord

ACTIVITY: A MODEL OF THE SPINAL CORD

You will need:

- Plasticine
- Single core, plastic insulated wire
- White beads
- Wire strippers

Method:

1. Make a Plasticine model of a cross-section of a spinal cord. Use a different colour for the central grey matter.

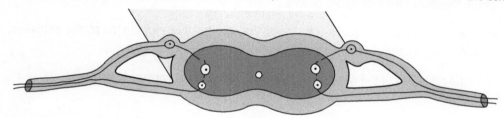

2. Make a sensory and motor neurone as described in activity 9.1.

3. Make an intermediate (connector) neurone. This only needs to be about 10 mm long, and doesn't have any insulation.

4. Arrange the neurones in the correct sequence on the model spinal cord. This now represents a single reflex arc.

Alternative method: if no Plasticine is available, you can draw out a cross-section of the spinal cord, and lay your three neurones on it.

> **Extension**
>
> Young people may become infected by a bacterium causing **meningitis**. Use the internet or a library book to find out what the **meninges** are, and describe why this infection is so dangerous.

Project ideas

9.3 Naming the parts of the body

Scientists use a very specialised language when they describe living organisms, and the structures that make them up. You may have come across some of these words if you have studied sports science, or if you are interested in anatomy. Doctors use the same terms when they are describing parts of the human body: this diagram shows a few of them!

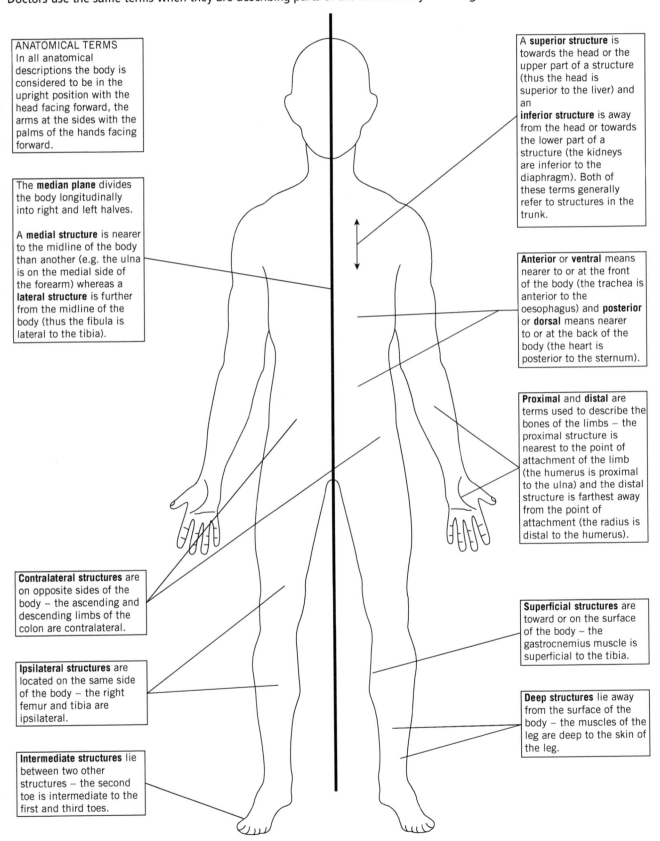

ANATOMICAL TERMS
In all anatomical descriptions the body is considered to be in the upright position with the head facing forward, the arms at the sides with the palms of the hands facing forward.

The **median plane** divides the body longitudinally into right and left halves.

A **medial structure** is nearer to the midline of the body than another (e.g. the ulna is on the medial side of the forearm) whereas a **lateral structure** is further from the midline of the body (thus the fibula is lateral to the tibia).

Contralateral structures are on opposite sides of the body – the ascending and descending limbs of the colon are contralateral.

Ipsilateral structures are located on the same side of the body – the right femur and tibia are ipsilateral.

Intermediate structures lie between two other structures – the second toe is intermediate to the first and third toes.

A **superior structure** is towards the head or the upper part of a structure (thus the head is superior to the liver) and an **inferior structure** is away from the head or towards the lower part of a structure (the kidneys are inferior to the diaphragm). Both of these terms generally refer to structures in the trunk.

Anterior or **ventral** means nearer to or at the front of the body (the trachea is anterior to the oesophagus) and **posterior** or **dorsal** means nearer to or at the back of the body (the heart is posterior to the sternum).

Proximal and **distal** are terms used to describe the bones of the limbs – the proximal structure is nearest to the point of attachment of the limb (the humerus is proximal to the ulna) and the distal structure is farthest away from the point of attachment (the radius is distal to the humerus).

Superficial structures are toward or on the surface of the body – the gastrocnemius muscle is superficial to the tibia.

Deep structures lie away from the surface of the body – the muscles of the leg are deep to the skin of the leg.

Project ideas   **9.3 Naming the parts of the body**

a. Look at the diagram on the previous page – don't attempt to learn the names, but try to answer these questions.

  i. Are the **medial** ligaments on the inside or the outside of the knee?

  ii. Is the big toe **proximal** or **distal** to the knee?

  iii. Is the navel (belly button) a **ventral** or a **dorsal** structure?

  iv. In a reflex arc, is the motor nerve **ventral** or **dorsal**?

  v. The left side of the brain controls muscles on the right side of the body – is this an **ipsilateral** or a **contralateral** relationship?

  Simpler animals follow similar rules, but anterior means 'head end' and posterior means 'tail end'. Dorsal still means 'back' (think of a shark's dorsal fin!) and ventral means 'front'.

b. Sportsmen and sportswomen are often at risk of injury in almost any sport, but particularly in contact games such as soccer and rugby.

  Use the diagram on the previous page, and any other resources you have access to, to explain:

  i. Why soccer players suffer from cruciate ligament injuries. Can you make a simple model to demonstrate this?

  ii. Why a blow with a cricket ball on the left side of the skull might cause numbness and even paralysis in the right side of the body.

  iii. Why the main vein returning to the heart from the body can be called the **inferior** or the **posterior** vena cava.

c. Make a drawing of an Annelid, such as an Earthworm. Add labels, using the anatomical terms from the diagram. Add a scale to your diagram.

Glossary

Absorption: Movement of small food molecules and ions through the wall of the intestine into the bloodstream.

Adaptive feature: An inherited feature that helps an organism to survive and reproduce in its environment.

Adaptive feature: The inherited functional features of an organism that increase its fitness.

Active immunity: Defence against a pathogen by antibody production in the body.

Active transport: The movement of particles from a region of lower concentration to one of higher concentration, across a membrane, using energy from respiration.

Aerobic respiration: The chemical reactions in cells that use oxygen to break down nutrient molecules to release energy.

Allele: A version of a gene.

Anaerobic respiration: The chemical reactions in cells that break down nutrient molecules to release energy without using oxygen.

Asexual reproduction: A process resulting in the production of genetically identical offspring from one parent.

Assimilation: The use of digested food molecules by cells.

Binomial system: Naming species in a system in which the scientific name of an organism is made up of two parts showing the genus and species.

Carnivore: An animal that gets its energy by eating other animals.

Catalyst: A substance that increases the rate of a chemical reaction and is not changed by the reaction.

Chemical digestion: Breakdown of large, insoluble food molecules into small, soluble molecules.

Chromosome: A thread-like structure of DNA, carrying genetic information in the form of genes.

Community: All of the populations of different species in an ecosystem.

Consumer: An organism that gets its energy by feeding on other organisms.

Cross-pollination: Transfer of pollen grains from the anther of a flower to the stigma of a flower on a different plant of the same species.

Deamination: The removal of the nitrogen-containing part of amino acids to form urea.

Decomposer: An organism that gets its energy from dead or waste organic material.

Diffusion: The net movement of particles from a region of higher concentration to one of lower concentration down a concentration gradient, as a result of their random movement.

Diploid nucleus: A nucleus containing two sets of chromosomes, e.g. in body cells.

Dominant: An allele that is expressed if it is present.

Drug: Any substance taken into the body that modifies or affects chemical reactions in the body.

Ecosystem: A unit containing the community of organisms and their environment, interacting together, e.g. a decomposing log, or a lake.

Egestion: The passing out of food that has not been digested or absorbed, as faeces through the anus.

Enzyme: A protein that functions as a biological catalyst.

Excretion: Removal of toxic materials and substances in excess of requirements.

Fertilisation: The fusion of gamete nuclei.

Fitness: The probability of an organism surviving and reproducing in the environment in which it is found.

Food chain: The transfer of energy from one organism to the next, beginning with a producer.

Food web: A network of interconnected food chains.

Gene: A length of DNA that codes for a protein.

Gene mutation: A change in the base sequence of DNA.

Genetic engineering: Changing the genetic material of an organism by removing, changing, or inserting individual genes.

Genotype: The genetic make-up of an organism in terms of the alleles present.

Gravitropism: A response in which parts of a plant grow towards or away from gravity.

Growth: A permanent increase in size and dry mass by an increase in cell size and/or cell number.

Haploid nucleus: A nucleus containing a single set of unpaired chromosomes, e.g. in gametes.

Herbivore: An animal that gets its energy by eating plants.

Heterozygous: Having two different alleles of a particular gene.

Homeostasis: The maintenance of a constant internal environment.

Homozygous: Having two identical alleles of a particular gene.

Hormone: A chemical substance, produced by a gland and carried by the blood, which alters the activity of one or more specific target organs.

Ingestion: The taking in of substances into the body through the mouth.

Glossary

Inheritance: The transmission of genetic information from generation to generation.

Limiting factor: Something present in the environment in such short supply that it restricts life processes.

Mechanical digestion: Breakdown of food into smaller pieces without chemical change to the food molecules.

Meiosis: Nuclear division giving rise to cells that are genetically different.

Mitosis: Nuclear division giving rise to genetically identical cells.

Movement: An action causing the change in position of an organism or part of an organism.

Mutation: Genetic change.

Nutrition: The taking in of materials for energy, growth, and development.

Organ: A structure made up of a group of tissues, working together to perform a specific function.

Organ system: A group of organs with related functions.

Osmosis: The net movement of water molecules from a region of higher water potential to a region of lower water potential, through a partially permeable membrane.

Passive immunity: Short-term defence against a pathogen by antibodies acquired from another individual, e.g. mother to infant.

Pathogen: A disease-causing organism.

Phenotype: The observable features of an organism.

Photosynthesis: The process by which plants manufacture carbohydrates from raw materials using energy from light.

Phototropism: A response in which parts of a plant grow towards or away from the direction from which light is coming.

Pollination: Transfer of pollen grains from anther to stigma.

Population: A group of organisms of one species, living in the same area, at the same time.

Process of adaptation: The process, resulting from natural selection, by which populations become more suited to their environment over many generations.

Producer: An organism that makes its own organic nutrients, usually using energy from sunlight, through photosynthesis.

Recessive: An allele that is only expressed when there is no dominant allele of the gene present.

Reproduction: The processes that make more of the same kind of organism.

Respiration: The chemical reactions in the cell that break down nutrient molecules to release energy for metabolism.

Self-pollination: The transfer of pollen grains from the anther of a flower to the stigma of the same flower or different flower on the same plant.

Sense organ: Groups of receptor cells responding to specific stimuli: light, sound, touch, temperature, and chemicals.

Sensitivity: The ability to detect and respond to changes in the environment.

Sex-linked characteristic: A characteristic in which the gene responsible is located on a sex chromosome.

Sexual reproduction: A process involving the fusion of the nuclei of two gametes (sex cells) to form a zygote and the production of offspring that are genetically different from each other.

Sexually transmitted infection: An infection that is transmitted via body fluids through sexual contact.

Species: A group of organisms that can reproduce to produce fertile offspring.

Sustainable development: Development providing for the needs of an increasing human population without harming the environment.

Sustainable resource: One which is produced as rapidly as it is removed from the environment so that it does not run out.

Synapse: A junction between two neurones.

Tissue: A group of cells with similar structures, working together to perform a shared function.

Translocation: The movement of sucrose and amino acids in the phloem.

Transmissible disease: A disease in which the pathogen can be passed from one host to another.

Transpiration: Loss of water by plant leaves by evaporation and diffusion.

Trophic level: The position of an organism in a food chain, food web, pyramid of numbers, or pyramid of biomass.

Variation: Differences between individuals of the same species.

Answers

Unit 1.1
1. A: sensitivity, B: reproduction, C: excretion, D: respiration
2. C
3. Kingdom – phylum – class – order – family – genus – species

Unit 1.2
1. a. *Parus caeruleus* and *Parus major* – they belong to the same genus
   b. i.

E. rubecula	×	×	×
P. caeruleus	×	✓	✓
P. major	×	✓	✓
T. merula	×	×	×

   ii. Pale area only below eye *Parus major*
   Pale areas above and below eye *Parus caeruleus*

Unit 1.3
1. a. Stem – hold leaves in best position; root – absorb water and mineral ions; leaves – trap light energy for photosynthesis; flowers – may be attractive to pollinating insects or birds; fruit – usually help dispersal of seed, a reproductive structure
   b. chlorophyll   autotrophic   photosynthesis   cellulose
      algae   ferns   angiosperms   monocotyledons   dicotyledons

Unit 1.4
1. a. Ant: three/yes; earthworm: none/no; centipede: many/yes; mite: four/no
   b. Ant – insect; earthworm – annelid; centipede – myriapod; mite – arachnid

Unit 1.5
1. a. An animal with a backbone
   b. Scales; no; no; yes; fur – yes

Unit 1.6
1. a. Phloem – transport; stamens – reproductive.
   b. i. Could label nucleus/cell membrane/cytoplasm
      ii. chloroplast
      iii. cellulose cell wall/permanent vacuole
   c. i. mitochondrion/mitochondria
      ii. ribosomes (accept rough endoplasmic reticulum)

Unit 1.7
1. a. From top line, reading right to left: cell C, cell B, cell F, cell I, cell E, cell A, cell H, cell D, cell G
   b. leaf

Unit 2.1
1. a. Diffusion   down   gas   random   equilibrium
      Osmosis   diffusion   potential   partially permeable

Unit 2.2
1. a. i. +8, +4, −1, −6, −10
      ii. graph plotted
      iii. 0.45
      iv. Represents an equilibrium so shows water potential of cell cytoplasm.
   b. osmosis

Unit 2.3
1. a. glucose   cellulose   sucrose   soluble
   b. fatty acids   glycerol   insoluble
   c. haemoglobin   amino acids   soluble
   d. DNA
2. Protein: + + + − − +; Fat: + + + − − −;
   Carbohydrate: + + + − − −; Nucleic acid: + + + − + +

Unit 2.4
1. a. i. Higher protein content.
      ii. Earlier step in food chain, so more can be produced per unit area. Beef cattle must live longer before they can be killed for meat.
   b. i. sample Z
      ii. as a reagent blank
      iii. sample X

Unit 2.5
1. Protein – the type of molecule that makes up an enzyme; substrate – a molecule that reacts in an enzyme-catalysed reaction; product – the molecule made in an enzyme-catalysed reaction; active site – the part of the enzyme where substrate molecules can bind; denaturation – a change in shape of an enzyme so that its active site cannot bind to the substrate; optimum – the ideal value of a factor, such as temperature, for an enzyme to work
2. Lipase; cuts out useful genes from chromosomes; removes milk sugar from milk; protease; cellulase; amylase

Unit 2.6
1. a. Check correct column headings e.g. temperature (°C) and time for clotting (minutes)
   b. Graph plotted – temperature on x axis
   c. 45 °C
   d. i. pH
      ii. Use buffer solutions

Unit 2.7
1. Photosynthesis   light/solar   chloroplasts/chlorophyll
   carbon dioxide   water   starch   oxygen   stomata
2. a. carbon dioxide
   b. oxygen
   c. water
   d. nitrate
   e. magnesium

Unit 2.8
1. a. 3.5 arbitrary units
   b. Light intensity is the limiting factor up to 6 arbitrary units. At this point, some other factor (temperature, for example) is at a value which is preventing further photosynthesis.

Unit 2.9
1. a. A – upper epidermis; B – palisade mesophyll; C – spongy mesophyll; D – guard cell; E – xylem vessel
   b. B
   c. E
   d. Sucrose
   e. i. 0.9 mm
      ii. 0.25 mm

Unit 2.10
1. a. Light intensity 10 arbitrary units, carbon dioxide concentration 0.15% at 30°C. Make sure that you give all of the units in your answer!
   b. Even at the same light intensity and temperature, increasing the carbon dioxide concentration from 0.04% to 0.15% increased the rate of photosynthesis by as much as 375%.
   c. Carbon dioxide concentration is the limiting factor under those sets of conditions.

Unit 2.11
1. a. To act as a control. In this case to show that the indicator does not change colour due to some other factor.
   b. To make sure that this is a fixed variable i.e. volume of indicator should not affect the results.
   c. i. photosynthesis
      ii. respiration
      iii. respiration
   d. These plants require a high light intensity to photosynthesise, and light intensity would be low in the forest environment.

Answers

Unit 2.12
1. a. i. To allow measurement of the maximum (complete solution) and minimum (water) rate of growth of the wheat seedlings.
      ii. Air contains oxygen and oxygen is needed for aerobic respiration. This releases energy for active uptake of mineral ions by the root hair cells.
      iii. Magnesium is required to produce chlorophyll. No chlorophyll means very limited photosynthesis so very poor growth.
   b. nitrogen   phosphorus   potassium

Unit 2.13
1. a. 70% = 55% + 15% (each segment represents 5%)
   b. From the top of the list: T; F; T; F; T; T; F; T; F; F; T; F; F; T; F

Unit 2.14
1. a. i. age (teenage boy needs more energy than eight-year-old); gender – male IT worker needs more energy than female IT worker; activity – manual worker needs more energy than IT worker
      ii. female is lighter than a male, so less mass to move about
      iii. Vitamin D: rickets (more likelihood of broken bones/night blindness). Iron – anaemia/tiredness, since less haemoglobin/red blood cells produced so less oxygen transported.

Unit 2.15
1. a. Carbohydrate and fat/lipid
   b. i. More
      ii. $\frac{2719}{9000} \times 100 = 30.2\%$
      iii. Jack is heavier/more active/growing more quickly

Unit 2.16
1. a. i. male
      ii. female, aged 20–24
   b. i. as fat
      ii. diabetes/arthritis/heart disease
   c. Provision of an unbalanced diet.
   d. i. protein required for growth/production of enzymes/haemoglobin
      ii. lethargy/poor rate of growth/inactivity

Unit 2.17
1. Salivary glands – produce an alkaline fluid which………..; oesophagus – carries a bolus of food from mouth to stomach; stomach – produces hydrochloric acid and begins digestion of protein; Ileum – where most digested food is absorbed; pancreas – produces a set of enzymes which pass into the duodenum; gall bladder – stores bile produced in the liver; colon – where most of the water is reabsorbed……; rectum – stores waste food as faeces.
2. bread   butter   egg   mouth   molars   surface area   large   smaller   capillaries

Unit 2.18
1. a. i. A – dentine; B – cement; C – crown.
      ii. blood vessels/nerves
   b. i. the enamel is so hard
      ii. nerves are inside the pulp cavity
   c. Vitamin C

Unit 2.19
1. a. With iodine solution: straw-brown   blue-black   blue-black   blue-black
      With Benedict's reagent: orange-red   blue   blue   blue
   b. Mouth (from salivary glands), small intestine (from pancreas)
   c. This is the optimum temperature for amylase activity
   d. In the stomach

Unit 2.20
1. a. Villus
   b. Small intestine/ileum
   c. i. M – Should be linked to lacteal (at centre of villus);
      ii. N – should be linked to surface epithelium (produces mucus from goblet cells)
   d. large surface area increases rate of absorption, thin epithelium reduces distance for absorption, capillaries/lacteals remove absorbed products.
   e. i. ×50
      ii. 1.5 mm

Unit 2.21
1. osmosis   hairs   ions   magnesium   diffusion   active transport   support   solvent   photosynthesis

Unit 2.22
1. a. A: phloem;   B: xylem;   C: epidermis
   b. i. B because this is xylem which transports water and water-soluble dyes such as eosin
      ii. Radioactive carbon dioxide is converted to radioactive carbohydrate by photosynthesis. The carbohydrate is transported, as sucrose, in the phloem. The phloem therefore shows up by fogging the photographic plate.

Unit 2.23
1. a. Mean mass at start = 220.4 g. Mean mass after 24 h = 210.2 g. Therefore mean loss of water = 220.4 – 210.2 = 10.2 g.
   b. Mean volume of water at start = 100 cm^3 = 100 g. Mean volume of water after 24 h = 88 cm^3 = 88 g. Therefore mean uptake of water = 12 g.
   c. Water uptake is driven by water loss from the stomata/leaves. Water lost 'pulls' a stream of water through the plant – this is replaced by water uptake at the roots.
   d. Some of the water absorbed by the roots is used for support or in photosynthesis, so the masses are not exactly the same.

Unit 2.24
1. a. D is incorrect
   b. This shape of leaf would mean much water loss in the dry desert environment.
2. a. Look for water vapour loss by diffusion through stomata, water evaporation from surface of spongy mesophyll cells.
   b. Windspeed/humidity of atmosphere/light intensity

Unit 2.25
1. a. i. 100 – 45 = 55%
      ii. ions /glucose/ amino acids/ hormones
      iii. Red blood cell – transport of oxygen; phagocyte – engulfing invading microbes; lymphocyte – antibody production; platelet – part of clotting process
   b. Severe breathlessness/tiredness/inability to train hard
   c. They will have an increased number of red blood cells so their blood will be able to supply more oxygen than normal to the muscles, thus improving their performance.
   d. Testosterone (or some other steroid) – used because it increases growth of muscle tissue; Anabolic steroid / muscle building; stimulants such as amphetamine / more powerful muscle contractions

Unit 2.26
1. a. i. left ventricle
      ii. thicker muscular walls
   b. pacemaker   72
2. double
3. a. Blood pressure falls as blood passes through the gills, and cannot be raised again before blood reaches body tissues.
   b. kidneys
   c. Blood at low pressure can flow more easily through the wider veins.
   d. pulmonary arteries   tissue fluid/plasma

Unit 2.27
1. a. Blood returns to the heart twice for each complete circuit of the body.
   b. i. Blood pressure falls, from 12 kPa to 4 kPa.
      ii. From plasma to tissue fluid: glucose/amino acids/oxygen/ions such as iron and calcium; from tissue fluid to plasma: carbon dioxide/lactic acid/urea.
   c. Blood pressure would fall (temporarily).
   d. Poor filtration of blood at kidneys so build-up of toxic waste.
   e. Large surface area/wall is only one cell thick.

Answers

Unit 2.28
1. a. Change the depth of heartbeat (stroke volume), i.e. volume of blood pumped with each heartbeat
   b. i. X - cuspid valve - pressure in ventricle exceeds pressure in atrium so valve closes to prevent backflow
      Y - semilunar valve closes, as pressure in left ventricle falls below pressure in aorta, to prevent backflow of blood into heart
      ii. In the veins
      iii. Maintain one-way flow of blood at lower pressure in the veins
   c. Blood flow to lungs is reduced so blood does not oxygenate well, and so blood (which shows up in the lips) is not as red as it normally would be.

Unit 2.29
1. a. i. From the top: 77, 72, 96, 81, 76, 102, 72, 89.
      ii. The boys appear to have a higher pulse rate.
   b. i. Exercise increases blood flow to heart, muscles, and skin, but reduces blood flow to gut and kidneys. Flow to brain does not change.
      ii. Blood delivers more oxygen and glucose for respiration needed to release the energy required for muscle contraction.
   c. i. Less oxygen is delivered to working heart muscle, so less energy is available and muscle stops working.
      ii. obesity/high fat consumption/high salt consumption

Unit 2.30
1. a. High temperature/vomiting/diarrhoea
   b. Not washing hands when preparing food for child/not using clean utensils and crockery/feeding with infected food
   c. Always wash hands when preparing food for child/cook food thoughly and at the correct temperature
2. Rickets/scurvy; cystic fibrosis/sickle cell anaemia; dementia/heart disease/cancer; lung cancer/CHD/diabetes type II

Unit 2.31
1. a. An organism capable of causing a disease.
   b. Cholera bacterium – in infected water; influenza virus – in droplets in the air; athlete's foot fungus – by direct contact; plasmodium protoctist – by an insect vector; salmonella bacterium – in contaminated food; HIV – in infected body fluids
2. In infected body fluids; fungus; drying skin carefully; in infected water; bacterium; disposal of faeces correctly; bacterium; careful food preparation/cooking food thoroughly; virus; trapping 'sneezes' and 'coughs' in tissues or a handkerchief.

Unit 2.32
1. a. Food source column: fish; milk products; fresh meat; vegetables
      Number of reported outbreaks column: 20; 45; 5
   b. At refrigeration temperatures, bacteria which might be present in the food cannot multiply, so cannot produce toxins which cause food poisoning.

Unit 2.33
1. a. A disease caused when cells divide/multiply out of control. The extra cells can cause damage directly or may use up nutrients from other 'normal' cells.
   b. i. breast cancer
      ii. $\frac{9}{100} \times 1000 = 90$
      iii. ovary, cervix, uterus
   c. The disease can be identified before it causes too much damage. Treatment may destroy the cancer cells and so prolong life.

Unit 2.34
1. Pathogen – a disease-causing organism; transmissible disease – condition in which the pathogen can be passed from one host to another; antigen – a substance which triggers the immune response; skin and nasal hairs – external barriers to infection; active immunity – defence against a pathogen by antibody production in the body; passive immunity – short-term defence by provision of antibodies from another individual; phagocyte – white blood cell that engulfs pathogens; lymphocyte – cell which produces antibodies

2. a. A chemical or cell which provokes the body's immune system to produce appropriate antibodies.
   b. Response is faster, greater, and lasts for longer.

Unit 2.35
1. a. Defence against a pathogen by the production of antibodies in the body.
   b. Antibodies are from another individual/the response does not last as long/no memory cells are produced.
      Active   passive   passive   passive   active

Unit 2.36

(Crossword answers)
- 5 across: ADENOSINE
- 6 across: HEAT
- 7 across: GROWTH
- 9 across: ACTIVE TRANSPORT
- 10 across: RESPIRATION
- 12 across: MITOCHONDRIA
- 13 across: OXIDATION
- 14 across: CELL DIVISION
- 15 across: MOVEMENT
- 16 across: TRIPHOSPHATE
- 18 across: GLUCOSE
- 19 across: PROTEIN
- 20 across: ANAEROBIC
- 22 across: AEROBIC
- 23 across: LIGHT
- 24 across: LIVER

Down clues include: SOLA, PO, CNT CAT I, E, GYROK, WORK, KI, NL, MJ, OL, YN, SYNTHESIS, OXYGEN

Unit 2.37
1. a. i. $C_6H_{12}O_6 + 6O_2 \rightarrow 6CO_2 + 6H_2O$ + energy
      ii. Respiration is catalysed by enzymes, which are affected by temperature.
   b. i. Pie chart B – most of the energy must be supplied by aerobic respiration as the athlete could not continue to run if anaerobic respiration was the energy supplier (much less efficient, and lactate is toxic).
      ii. Performance is reduced as the lactic acid stops impulses reaching the muscle cells.
      iii. The lactic acid is removed by diffusing into the blood.
   c. The extra oxygen required to remove lactic acid accumulated during anaerobic exercise.

Unit 2.38
1. a. i. aerobic respiration
      ii. Towards the glass tube with the fruits in it.
      iii. The fruits use up oxygen, and the carbon dioxide they release is absorbed by the soda lime. Therefore the volume of gas inside the tube falls.
      iv. $\frac{40}{5} = 8.8 \times 0.25 = 2$ cm.
          Therefore the marker drop will be at 3.25 on the scale (i.e. current position is 5.25).
      v. Suitable control would be the tube as set up exactly the same but without the tomato fruits.
   b. Nitrogen gas does not allow respiration so that the fruits will not over-ripen and become 'spoiled'.

Unit 2.39
1. a. Oxygen from alveolus to blood, and carbon dioxide from blood to alveolus.
   b. large surface area/moist lining/very thin walls/closeness to capillary system
   c. pulmonary artery

Answers

Unit 2.40
1. a. lung
   b. Label a line drawn across the thorax below the lung.
   c. external intercostal muscles
   d. Intercostal muscles relax so front of thorax falls. Diaphragm relaxes so bottom of the thorax rises. Lung volume falls, so air is pushed out (exhaled).

Unit 2.41
1. a. i. Effect of age on the likelihood of COPD.
      ii. Chronic obstructive pulmonary disorder
      iii. No – there is no information about smoking, only about the age of sufferers from COPD.
   b. Vital capacity would fall. This is because smoking damages the alveoli so that they become less elastic. As a result they cannot stretch so the vital capacity is reduced.

Unit 2.42
1. If carbon monoxide combines with the haemoglobin, the red blood cells cannot carry so much oxygen. The developing fetus requires oxygen from its mother's blood, so will suffer if the mother has less oxygen in her red cells.
2. a. Smoker of 20 cigarettes per day has a risk factor of 2.3, a non-smoker has a risk factor of 0.1. Therefore a smoker is about 23 times more likely to die from lung cancer.
   b. Smoker of 1-14 cigarettes a day is 13 times more likely to die of lung cancer than a non-smoker.

Unit 2.43
1. a. ammonia
   b. Optimum temperature for the enzyme urease.
   c. i. protein and glucose
      ii. They are reabsorbed in the kidney tubule.
   d. It is a control – to show that distilled water does not affect the biuret or Benedict's reagents.

Unit 2.44
1. a. This increases the surface area for diffusion.
   b. This will slightly raise the filtration pressure as the blood enters the dialysis machine.
   c. i. Plasma would have higher than normal concentrations of urea, chloride, and sodium ions.
      ii. ammonia/glucose/urea
      iii. higher/same; same; same; lower; lower; lower

Unit 2.45
1. a. A: sweat duct/pore;   B: dermis;   C: branch of vein
   b. They act as insulation – to retain heat within the body.
   c. Optimum temperature for enzymes/cell membranes do not break down/solutes remain dissolved in plasma

Unit 2.46
1. a. A: sensory neurone; B: hair shaft
   b. i. It is raised.
      ii. Keeps a layer of air close to the skin, acting as a thermal insulator.
      iii. capillary constricted, position the same
   c. This is a control system in which a deviation from the norm sets up a series of changes which cancel out the deviation.

Unit 2.47
1. From the top of the table: A; C; E/F; E; D; B

Unit 2.48
1. a. i. Nose
      ii. They always have a survival value e.g. clearing the eye of particles of dust.
      iii. rapid/involuntary/short-lived/always have survival value
   b. i. 1 – receptor   2 – sensory neurone   3 – white matter
         4 – grey matter   5 – association neurone   6 – synapse
         7 – motor neurone
      ii. muscles – contract glands – secrete (e.g. a hormone)

Unit 2.49
1. a. 6 – 9 – 9 – 4 – 2
   b. i. 25s
      ii. 50%
   c. At 60 km per h car travels 1km in 1 minute so $\frac{1000}{60}$ metres in 1 second. Slowest response is 0.37s i.e. after $\frac{1000}{60} \times 0.37 = 6.2$ m

Unit 2.50
1. a. From the top of the table: C; F; B; A; E; D
   b. i. stimulus is bright light; receptor is retina; co-ordinator is central nervous system; effector is iris; response is reduce diameter of pupil
      ii. Prevents bright light from damaging the retina.

Unit 2.51
1. a. Hormone: a chemical, produced by an endocrine organ, released into the bloodstream where it brings about a response at a target organ.
      Target organ: an organ which brings about a response to stimulation by a specific hormone.
   b. Hormones: slower/longer-lasting/more general than nervous control
2. a. Adrenaline is released – this stimulates the conversion of stored glycogen into soluble glucose.
   b. 1. Glucose must be absorbed from the gut – this takes a finite amount of time.
      2. Pupils dilate/skin becomes pale/hair stands on end/heart beats more quickly.

Unit 2.52
1. a. From top of table: C; A; B; E/F
   b. Alcohol transported in maternal blood/diffuses across placenta/enters fetal bloodstream
2. Heroin – feeling of calm and rest; nicotine – stimulates the heart rate; oestrogen – stimulates ovulation; penicillin – protects against syphilis; testosterone – increases growth rate in muscle

Unit 2.53
1. a. i. Auxin has moved downwards and to the non-illuminated side of the shoot.
      ii. Auxin moves to non-illuminated side – causes enlargement of cells on this side – shoot therefore bends towards the light.
      iii. Name of response – (positive) phototropism; Benefit – moves leaves into the optimum position for photosynthesis
   b. Co-ordinating formation of fruit for easier harvesting/acting as a weedkiller in growing crops

Unit 3.1
1. a. mitosis in the asexual diagram, meiosis in the sexual diagram.
   b. for asexual   nn   and   nn;   for sexual   n   and   n
   c. advantage : rapid/progeny well adapted to same environment as parents
      disadvantage : no variation possible, so great risk if 'unknown' infection attacks population

Unit 3.2
1. a. Transfer of pollen from anther to the stigma.
   b. Z: petal;   W: anther;   Y: stigma;   X: ovary
   c. Petals: small and dull-coloured;   large and brightly coloured; attraction of insects
      Anthers: hang out into the wind;   kept deep inside flower; anthers can release pollen into the wind
      Pollen: very light;   heavy and produced in smaller quantities; light pollen can float in the wind
      Stigma: branched;   deep within flower, strong;   branched to catch wind-blown pollen/strong to avoid damage from visiting insect

Answers

Unit 3.3
1. a. i. Mark P on tip of stigma
      ii. stigma
   b. i. Show entry via the micropyle
      ii. Male nucleus fuses with female nucleus to form a diploid zygote. Permits variation by random combination of genetic material from parent plants.

Unit 3.4
1. a. From the top of the column: E; C; B; A; D; F
   b. Show pollen tube growth down style and entry to ovule via the micropyle
   c. i. plumule and radicle
      ii. starch
      iii. Iodine solution has changed from straw-brown to blue-black.

Unit 3.5
1. a. From left to right of bottom row: yes – no – yes – no – no – no
   b. Suitable temperature (for enzyme action)/water/presence of oxygen/unaffected by the presence or absence of light

Unit 3.6
1. a. They digest the coating of the female gamete to allow the head of the sperm to enter.
   b. Release energy by aerobic respiration – energy needed for beating of tail.
   c. $50\,\mu m$
   d. The sperm is haploid (just like the egg cell) and most body cells are diploid.
   e. Site: testis – ovary; Numbers: millions – few; Mobility: mobile – unable to move itself; Relative size: smaller – larger

Unit 3.7
1. a. i. 8
      ii. 2nd July
      iii. No egg cell is available.
   b. i. progesterone
      ii. oestrogen

Unit 3.8
1. a. From top of diagram: oviduct; ovary; uterus; vagina
   b. i. in oviduct/uterus
      ii. zygote
   c. From top of column: fertilisation; conception; copulation; AID; implantation; development

Unit 3.9
1. a. Prevents conception/prevents transmission of infected body fluids
   b. i. From the top of the column: 5 – 3 – 4 – 2 – 6 – 1
      ii. Combination of progesterone and oestrogen reduces the chance of ovulation (feedback inhibition). No ova available means no possibility of fertilisation/conception.
      iii. $1000 \times \frac{1}{20} = 50$.

Unit 3.10
1. a. i. amniotic cavity – contains amniotic fluid which acts as a shock absorber; uterus wall – muscular and so is protection against physical damage
      ii. from end of umbilical cord to wall of uterus
   b. Glucose – from mother to fetus; haemoglobin – no movement across placenta; nicotine – from mother to fetus; amino acids – from mother to fetus; carbon dioxide – from fetus to mother; alcohol – from mother to fetus; urea – from fetus to mother

Unit 3.11
1. a. $\frac{2400}{8800} \times 100 = 27\%$
   b. More calcium and vitamin D for rapid growth of developing bones; less iron because the infant has less need to replace haemoglobin as it has a lower blood volume.
   c. $\frac{2400}{2750} = 0.872 \times 1\,kg = 872\,g$.
   d. Cow's milk has too much protein, not enough iron or vitamin D.
   e. Vitamin D is synthesised in the skin when the skin is exposed to sunlight.

Unit 3.12
1. progesterone falls oxytocin oestrogen rises
2. a. i. gestation period
      ii. 38 weeks
   b. i. Head is the largest part so once it is out of the uterus the body can easily follow.
      ii. Contraction of the muscular walls of the uterus.

Unit 3.13
1. a. i. bacterium
      ii. painful urination
      iii. condom
   b. i. Reduces the production of defensive white blood cells.
      ii. AIDS is viral, and so is not affected by antibiotics, which can control the bacterium causing gonorrhoea.

Unit 3.14
1. a. i. gender/eye colour/blood group
      ii. An alternative form of a gene.
      iii. show a number of genes lined up along a thread-like chromosome
   b. i. 9
      ii. could be Rr or RR
      iii. $\frac{3}{4}$ or 75%

Unit 3.15
1. a. GCCTATG
   b. i. messenger RNA
      ii. ribosome
   c. amino acid
   d. i. protein
      ii. haemoglobin - antibodies - can recognise and bind to a neurotransmitter - enzyme which breaks down fats to fatty acids and glycerol - keratin

Unit 3.16
1. a. amylase
   b. Advantages: protein can be made more cheaply/it can be purer/it can be produced when required
   Concerns: modified organisms might escape into the environment/ superweeds could be created if modified plants pollinate wild plants/genetically modified products could be expensive if the organisms are patented by drug companies

Unit 3.17
1. a. Male nerve cell: 46; XY; female white blood cell: 46; XX; sperm cell: 23; X or Y; egg cell: 23; X; red blood cell: 0; none
   b. mitosis
2. a. 4
   b. mitosis
   c. bone marrow/liver/skin/apical meristem of plant

Unit 3.18
1. gene meiosis haploid fertilisation diploid recessive heterozygous
2. Genotype – the set of alleles present in an organism; homozygous – having two identical alleles; dominant – an allele that is always expressed if it is present; heterozygous – having two alternative alleles; recessive – allele that is only expressed in a homozygous individual; chromosome – a thread-like structure of DNA…; allele – one alternative form of a gene; phenotype – the observable features of an organism

Unit 3.19
1. a. i. Andrew and John
      ii. Blood group – a discontinuous variation, so affected by genes only.
      iii. Nutrition – David might eat more high-calorie foods, or take less exercise.
   b. co-dominance

Answers

Unit 3.20
1. a. Dominant: N; Recessive: n
   b. Philip: Nn; Janet: Nn
   c. i. nn
      ii. Nn or NN
   d. $\frac{1}{4}$ or 25%
   e. Pancreatic duct can be blocked by sticky mucus, so these enzymes cannot enter the duodenum and the foods are not digested.

Unit 3.21
1. a. female, has 2 X chromosomes
   b. An extra 21st chromosome.
2. a. 2 has 2 X chromosomes, 3 has two X chromosomes, 5 has one X and one Y chromosome
   b. 2 could have H and h, 3 could have h and h, 5 could be h
   c. female – female – female – male – male

Unit 3.22
1. Continuous variation – a form of variation with many intermediate forms between the extremes; gene – a section of DNA responsible for an inherited characteristic; discontinuous variation – a form of variation with clear-cut differences between groups; phenotype – the observable features of an organism; height in humans – one example of continuous variation; environment – this factor, in addition to genotype, can affect phenotype; nutrients – one possible form of environmental influence on variation; blood group – one example of discontinuous variation.

Unit 3.23
1. a. i. Hb_S Hb_A
      ii. Tom, since both parents are heterozygous (carriers of the abnormal allele)
   b. This allele offers some protection in areas where malaria is common. The malarial parasite cannot reproduce so easily inside the red blood cells of a sickle cell individual.
2. a. i. Organisms with so many common features that they may interbreed and produce fertile offspring.
      ii. A change in the type or quantity of DNA in an individual.
      iii. A section of DNA responsible for coding for a single protein/characteristic.
   b. i. radiation
      ii. carcinogenic chemicals such as tar in cigarette smoke

Unit 3.24
1. a. artificial – artificial – natural – natural – natural
   b. i. By selecting individual animals with a high milk yield, breeding them, then selecting from their offspring cows that have a high milk yield, then breeding them, and so on.
      ii. resistance to disease/ability to withstand difficult environmental conditions such as low temperatures

Unit 3.25
1. a. B - E - C - A - D
   b. Heron – feeds by spearing fish and frogs; hawk – captures Florida rabbits and other mammals; spoonbill – filters algae and other small organisms from the water; finch – feeds on nuts and other hard fruits; warbler – feeds by catching small insects

Unit 3.26
1. a. Brussels sprouts – bud; broccoli – flower
   b. i. They have been produced by mitosis, so there is very little possibility of genetic variation.
      ii. nitrate – the ion needed for protein synthesis (growth); plants in closed environment – temperature and humidity can be controlled so that cuttings do not dry out

Unit 4.1
1. a. $0.5 \times 0.5 = 0.25$ m²
   b. There is a total of 60 springtails in 25 quadrats = 2.4 springtails per 0.25 m². Therefore there is a mean number of $2.4 \times 4 = 9.6$ springtails per m².
   c. They will not dry out/hidden from predators/can find rotting vegetation for food

Unit 4.2
1. Food chain – the transfer of energy from one organism to the next, beginning with a producer; food web – a network of interconnected food chains; producer – an organism that makes its own organic nutrients, usually through photosynthesis; consumer – an organism that gets its energy by feeding on other organisms; herbivore – an animal that gets its energy from eating plants; carnivore – an animal that gets its energy by eating other animals; decomposer – an organism that gets its energy from dead or waste material
2. Pyramid should have broad base, of algae, then gradually narrowing through water fleas, smelt, and to the pointed peak at kingfisher.

Unit 4.3
1. a. i. the Sun
      ii. feeding
      iii. some is reflected/the wrong wavelength/does not fall on leaves
      iv. respiration – energy is lost here as heat
   b. i. A: $\frac{15000}{90000} \times \frac{100}{1} = 16.6\%$; B: $\frac{2000}{15000} \times \frac{100}{1} = 13.3\%$
      ii. Each stage in the food chain allows a loss of energy as heat. Fewer steps, as in eating a vegetarian diet, mean that more energy is transferred with this heat loss.

Unit 4.4
1. a. i. C: 28; D: 17
      ii. For C (greatest loss in mass) environment is moist to provide water and large holes in mesh bag allow entry of oxygen.
      iii. C has larger holes so more oxygen can diffuse into the bag.
      iv. So that the bags themselves did not decompose.
   b. fungi and bacteria

Unit 4.5
1. a. From left to right and top to bottom: photosynthesis; respiration; feeding; decay
   b. It would raise the carbon dioxide concentration in the air.

Unit 4.6
1. a. i. B – feeding; E – denitrification
      ii. decomposition/decay by bacteria and fungi
   b. i. ammonium concentration would rise/nitrite and nitrate levels would fall/fewer dead leaves would be decomposed
      ii. ticks in few scavenging insects/ regular turning of the heap to add air

Unit 4.7
1. a. From left to right: rainfall; transpiration; condensation; evaporation; melting/freezing
   b. i. in urine/in breath/as sweat
      ii. the animal may eat the plant/the water is released from the plant cells/water absorbed by animal from gut/water transported in blood/water enters animal cells in tissues by osmosis

Unit 4.8
1. a. higher death rate/less food/lower birth rate/ emigration
   b. Vaccination/immunisation programmes; use of antibiotics; improved surgical techniques; fewer deaths at childbirth
   c. Use of antibiotics/antiseptics; thorough cooking to reduce food poisoning bacteria

Unit 4.9
1. a. The percentages of the different age groups that make up the population.
   b. 1: food availability – poor diet can lead to high death rate in younger age groups; 2: hygiene – food poisoning can reduce populations, especially in the very young and the elderly; 3: medical provision – immunisation programmes reduce death rates in lower age groups

Unit 4.10
1. a. From the top of the table: F – F – T – T – F – T – T – T – T – T – T – F
   b. Typical bacterium is one millionth of a metre wide/bacteria are larger than viruses/many bacteria are not harmful, they are useful/bacteria do not have a nucleus – they have 'naked' DNA

Answers

Units 4.11 and 4.12
1. a. Sugar is the raw material for respiration, which provides the energy for bacterial multiplication.
   b. To maintain the optimum temperature for the activity of bacterial enzymes, and to avoid damage to protein products.
   c. (6–) 7 days: this represents the maximum yield of penicillin – any time longer produces less penicillin but would be expensive in energy and raw materials.
   d. Some bacteria in the population have a natural resistance – the non-resistant bacteria are killed by the antibiotic – the resistant bacteria now multiply and produce a resistant population.
   e. Pectinase; protease; removes milk sugar from milk

Unit 4.13
1. a. i. To carry out respiration and release carbon dioxide. If water was substituted for sugar, no respiration could occur.
      ii. independent – type of sugar; dependent – volume of carbon dioxide produced
      iii. So that an accurate measurement of carbon dioxide volume could be made.
      iv. The concentration of sugar has fallen – it has become the limiting factor.
   b. glucose → ethanol + carbon dioxide
   c. Increase the temperature/increase the concentration of sugar/use purified enzymes from yeast

Unit 4.14
1. a. Gene – a section of DNA coding for a protein; plasmid – a small circle of DNA in a bacterial cell; vector – a structure which can carry a gene into another cell; ligase – an enzyme that can splice one gene into another section of DNA; restriction – an enzyme that can cut a specific gene from a chromosome; sticky ends – pieces of single-stranded DNA left exposed after a gene is cut from a chromosome; fermenter – a vessel in which engineered bacteria can produce a valuable product under optimum conditions.
   b. Insulin – control of blood sugar level; factor 8 – one step in blood clotting; pectinase – 'clears' fruit juices by breaking up clumps of plant tissue; human growth hormone – can increase growth rate in humans of short stature.
   c. The virus might not be completely inactivated, and thus may infect a patient with hepatitis.

Unit 4.15
1. a. 100 – 3 = 97%
   b. Each stage in the food chain allows a loss of energy as heat. Fewer steps, as in eating a vegetarian diet, mean that more energy is transferred with this heat loss.
   c. Insecticides may be present in the bodies of insects eaten by birds. The birds then accumulate the insecticide by eating many insects – this may harm the birds, e.g. affect egg laying.

Unit 4.16
1. Loss of habitat for animals/possible soil erosion/changes to water cycle/possible loss of valuable medicinal plants
2. a. i. Marsh has been drained/hedges have been removed/there are more buildings in 2003.
      ii. Marsh animals would be reduced in number e.g. frogs need water in which to lay their eggs; hedge removal would mean fewer nesting sites for birds; hedge removal would eliminate certain food plants for insects such as butterflies.
   b. Loss of nesting sites/soil erosion/loss of food plants for some species/less removal of carbon dioxide from the atmosphere/disturbance of water cycle.

Unit 4.17
1. a. i. beans
      ii. rice
      iii. vitamins and minerals
   b. Drought/loss of nutrients from soil/displacement of populations by war.

Unit 4.18
1. a. The gases form a layer around the Earth which allows radiation in but does not allow thermal radiation out – the surface of the Earth therefore becomes warmer.
   b. 1. These renewable energy resources do not produce greenhouse gases.
      2. Insulation means less heat loss so less combustion of fossil fuels for heating.
      3. Reforestation increases the number of plants which can remove carbon dioxide from the atmosphere during photosynthesis.
2. Removal of habitat for road construction / disturbance of wildlife by noise and presence of humans / pollution from vehicle exhausts

Unit 4.19
1. a. C - E - B - D - A
   b. Run-off of fertilisers from nearby farmland.
   c. Bacteria respire aerobically, using dead plants as food source. Aerobic respiration by bacteria reduces oxygen concentration in the water.
   Fish and larger invertebrates die as they are short of oxygen.

Unit 4.20
1. Management of the environment to maintain biodiversity.
2. a. i. Environmental factor: removal of bamboo forests – bamboo shoots are the basic food of the Giant Panda
      Biological factor: pandas reproduce very slowly – number of births does not replace number of dying pandas, so population falls
      ii. Saving the Panda (a very attractive species which many people will help to conserve) also protects other species in the same habitat.
   b. i. To find the population size at the start of the management plan.
      ii. Many examples possible e.g. butterfly – with nets, beetles – with pitfall traps, plants – with quadrats.

Unit 4.21
1. More shoals of fish are located, and more are caught by the fine mesh. Few fish escape to breed, so the population decreases. Loss of one species will affect the numbers of other species in the food web for this habitat.
2. Increase size of mesh/restrict areas in which fishing can take place/restrict times of year in which fishing can take place/limit the size of the catch for each fishing vessel.

Unit 4.22
1. a. A process which does not affect the environment for future generations.
   b. Less competition for light/for nutrients/for water.
   c. Limited food variety for animals/any pest or disease can affect all of the plants.
   d. Any will be a benefit. Best might be fast growth (so product is available quickly), resistance to disease (so few plants are lost to disease) and good growth on poor soil (so few fertilisers are necessary).

Unit 4.23
1. a. In the absence of oxygen.
   b. i. carbon dioxide
      ii. as dry ice for cooling foods in cold stores
   c. To avoid growth of potentially harmful microbes in the lake – the lake water might be consumed by humans.
   d. As a fertiliser – it has high concentrations of nitrate and phosphate.
   e. Disinfectants could kill the beneficial bacteria which decompose the sewage.

Unit 4.24
1. a. carbon dioxide + water → glucose + oxygen
   This process is driven by light energy
   b. The biocoil must be transparent to allow light to penetrate to the algae.
   c. The algae are removed to use as a fuel – they must be replaced quickly if the process is to be economical.

Answers

Unit 4.25
1. a. i. photosynthesis
      ii. diffusion
      iii. it is raised
   b. i. They are easier to decompose/ they are unlikely to harm wildlife which might consume them.
      ii. printing newspapers/as packaging/as toilet paper/as paper towels

Maths for Biology
Unit 6.1
1. i. 10
   ii. 1000
   a. A – cell surface membrane; B – cytoplasm; C – nucleus
   b. nucleus
   c. cellulose cell wall/chloroplast/permanent vacuole
   d. 3 cm = 30 000 µm; Magnification = $\frac{30\,000}{25}$ = ×1200

Unit 6.2
1. a. age group on x axis, frequency on y axis
   b. Increasing age leads to increasing frequency of multiple births (give some numerical examples to support your answer).

Unit 6.3
1. a. temperature on x axis, time taken on y axis
   b. 35°C
   c. These are fixed variables.
   d. Enzyme is denatured – loss of active site so cannot have binding of substrate to enzyme
   e. pH/enzyme concentration/substrate concentration.

Unit 6.4
1. a. 0.63 – 0.42 = 0.21; $\frac{0.21}{0.42} \times \frac{100}{1}$ = 50%
   b. Releasing carbon dioxide from cylinders of compressed gas gives good control of carbon dioxide concentration.

Unit 6.5
1. a. Using a pH meter on a sample of saliva collected (e.g. by chewing a rubber band).
   b. pH affects the number of fillings
   c. The increased pH shows a reduction in the number of fillings.

Unit 6.6
1. a. Time on x axis, mass on y axis. Two separate plots, with a key.
   b. Surface area will affect water loss. Keep this as a fixed variable.
   c. i. 4.35 g
      ii. loss = 0.8 g in two hours = 0.4 g per hour.

Unit 6.7
1. a. 8
   b. complete pie chart – largest sector beginning at '12'o'clock'
   c. Education to make people aware of the dangers of smoking/reduce passive smoking by banning smoking in public places.

Unit 6.8
1. a. independent – type of fruit; dependent – rate of respiration; fixed – temperature/mass of fruit/volume of air available
   b. Type of fruit; Experiment 1; Experiment 2; Experiment 3; Mean rate of respiration of fruit (mm³ used per min per g of fruit).
   c. Type of fruit on x axis, rate of respiration on y axis, bar chart, points plotted correctly.

Unit 6.9
1. a. 2 – 10 – 22 – 15 – 1
   b. number of seeds in pod on x axis, number of pods on y axis, bars should be touching
   c. 6
   d. $\frac{22}{50}$ × 100 = 44%

Unit 6.10
1. a. 1 – 2 – 4 – 6 – 10 – 5 – 3 – 2 – 1
   b. mass category on x axis, number in group on y axis, bars should be touching
   c. Continuous variation
   d. A result of a combination of genotype and effects of the environment.

Unit 6.11
1. a. time on x axis, number of bacteria on y axis
   b. sugar – energy from respiration; amino acids – protein synthesis; sterile – prevent competition from other bacterial species.
   c. show steepest part of curve
   d. shortage of oxygen/nutrients e.g. glucose or amino acids

Unit 6.12
1. a. Increase in population over time/ change in rate of increase from about 1920 onwards.
   b. 1960
   c. 1.2 billion
   d. Population at 1800 = 1 billion. Population reaches 2 billion in 1940, so time taken to double = 90 years.
   e. Approx. 9 billion

Exam-style questions
Unit 8.0
1. C: cell wall
2. Organism B
3. D
4. A
5. C
6. B
7. B
8. D
9. C
10. C
11. a. Correct position above and to the left of the stomach
    b. peristalsis
    c. From the top of the column: A – A – C – E – F
12. a. i. A: respiration; B: photosynthesis C: feeding D: respiration.
       ii. Decay/decomposition – a process in which organisms secrete enzymes which break down complex organic compounds such as proteins into simpler compounds such as amino acids/ammonium nitrate.
    b. i. In compost heap respiration is taking place. This releases energy, some of which is lost as heat.
       ii. This allows oxygen to penetrate the compost – oxygen is needed for aerobic respiration.
13. a. i. large surface area/close to blood supply/moist lining
       ii. Allows stretching during the inhalation/exhalation cycle.
    b. Blood at A has more carbon dioxide and less oxygen than blood at B.
    c. Makes lung tissue less elastic/breaks down walls of alveoli/inactivates cilia in airways/may trigger development of cancer cells.
14. a. i. A change in the quantity or type of DNA.
       ii. cystic fibrosis/Huntington's disease/albinism/sickle cell anaemia
    b. i. A recessive allele is only expressed in a homozygous individual.
       ii. A female affected with phenylketonuria.
       iii. individual 4: Pp; individual 6: PP or Pp
       iv. Individual 10 must be heterozygous i.e. Pp. Thus child has 1 in 4 chance of being homozygous i.e. showing phenylketonuria.
15. a. Temperature affects the activity of the enzymes which catalyse respiration, so experiment B will have enzymes working less efficiently.
    b. i. To prevent diffusion of oxygen from the air (keep respiration anaerobic).
       ii. Boiling removes oxygen from the solution.
    c. $C_6H_{12}O_6 \rightarrow 2C_2H_5OH + 2CO_2$
16. a. i. alleles
       ii. Parents are H_AH_S so can produce gametes H_A and H_S. The child could receive the mutant allele from both parents, and thus have the genotype H_SH_S.
       iii. 25% or 1/4
    b. i. The sickle cell allele provides protection against malaria, as the malaria parasite does not reproduce well inside sickle cells.
       ii. The medical treatment of malaria is advanced enough to prevent deaths of malaria sufferers so the sickle cell allele is not selected so positively.

Data sheet

You will find that the practical paper or the alternative to practical will be much more straightforward if you recognise certain standard pieces of equipment, and understand what they are used for.

Apparatus and materials

Safety equipment appropriate to the work being planned, but at least including eye protection such as safety spectacles or goggles.

3D image	Name and function	Diagram
	■ **Watch glass**: used for collection and evaporating liquids with no heat, and for immersing biological specimens in a liquid.	
	■ **Filter funnel**: used to separate solids from liquids, using a filter paper.	
	■ **Measuring cylinder**: used for measuring the volume of liquids.	
	■ **Thermometer**: used to measure temperature.	

The other very important measuring device in the laboratory is a balance (weighing machine).

	■ **Spatula**: used for handling solid chemicals; for example, when adding a solid to a liquid.	
	■ **Pipette**: used to measure and transfer small volumes of liquids.	
	■ **Stand, boss, and clamp**: used to support the apparatus in place. This reduces the risk of dangerous spills. This is not generally drawn. If the clamp is merely to support a piece of apparatus, it is usually represented by two crosses as shown.	
	■ **Bunsen burner**: used to heat the contents of other apparatus (e.g. a liquid in a test tube) or for **directly** heating solids.	HEAT
	■ **Tripod**: used to support apparatus above a Bunsen burner. **The Bunsen burner, tripod, and gauze are the most common way of heating materials in school science laboratories.**	
	■ **Gauze**: used to spread out the heat from a Bunsen burner and to support the apparatus on a tripod.	
	■ **Test tube and boiling tube**: used for heating solids and liquids. They are also used to hold chemicals while other substances are added and mixed. They need to be put safely in a test tube rack.	
	■ **Evaporating dish**: used to collect and evaporate liquids with or without heating.	
	■ **Beaker**: used for mixing solutions and for heating liquids.	

163

Data sheet

Table of hazard symbols

Symbol	Description	Examples
	Oxidising These substances provide oxygen which allows other materials to burn more fiercely.	Bleach, sodium chlorate, potassium nitrate
	Highly flammable These substances easily catch fire.	Ethanol, petrol, acetone
	Toxic These substances can cause death. They may have their effects when swallowed or breathed in or absorbed through the skin.	Mercury, copper sulfate

Symbol	Description	Examples
	Harmful These substances are similar to toxic substances but less dangerous.	Dilute acids and alkalis
	Corrosive These substances attack and destroy living tissues, including eyes and skin.	Concentrated acids and akalis
	Irritant These substances are not corrosive but can cause reddening or blistering of the skin.	Ammonia, dilute acids and alkalis

Reagent	Use in biology
hydrogencarbonate indicator (bicarbonate indicator)	Detects changes in carbon dioxide concentration, for example in exhaled air following respiration
✘ iodine in potassium iodide solution (iodine solution)	Detection of starch
✘ Benedict's solution (or an alternative such as Fehling's)	Detection of a reducing sugar, such as glucose
biuret reagent(s) (sodium or potassium hydroxide solution and copper sulfate solution)	Detection of protein
ethanol/methylated spirit	For dissolving lipids in testing for the presence of lipids. Also used to remove chlorophyll from leaves during starch testing.
cobalt chloride paper	Detects changes in water content
pH indicator paper or Universal Indicator solution or pH probes	Detects changes in pH during reactions such as digestion of fats to fatty acids and glycerol
litmus paper	Qualitative detection of pH
glucose	Change in water potential of solutions
sodium chloride	Change in water potential of solutions
aluminium foil or black paper	Foil can be used as a heat reflector: black paper as a light absorber
a source of distilled or deionised water	Change in water potential of solutions. Also used in making up mineral nutrient solutions.
eosin/red ink	To follow the pathway of water absorbed by plants
limewater	A liquid absorbent for carbon dioxide, for example in exhaled air
✘ methylene blue	A stain for animal cells
potassium hydroxide	Removes carbon dioxide from the atmosphere, for example during experiments on conditions for photosythesis
sodium hydrogencarbonate (sodium bicarbonate)	Very mild alkali: used for adjusting pH of solutions for enzyme activity
vaseline/petroleum jelly (or similar)	Blocks pores such as stomata, and so prevents water loss